Support Vector Machines for Antenna Array Processing and Electromagnetics

Support Vector Machines for Antenna Array Processing and Electromagnetics

Manel Mart´ınez-Ram´on and Christos Cristodoulou

ISBN:
978-3-031-00564-0   paperback
978-3-031-01692-9   ebook

DOI: 10.1007/978-3-031-01692-9

A Publication in the Springer series
SYNTHESIS LECTURES ON COMPUTATIONAL ELECTROMAGNETICS LECTURE #5
Series Editor: Constantine A. Balanis, Arizona State University

Series ISSN:
Print 1932-1252   Electronic 1932-1716

First Edition
10 9 8 7 6 5 4 3 2 1

Support Vector Machines for Antenna Array Processing and Electromagnetics

Manel Martínez-Ramón
Universidad Carlos III de Madrid, Spain

Christos Christodoulou
University of New Mexico

SYNTHESIS LECTURES ON COMPUTATIONAL ELECTROMAGNETICS #5

ABSTRACT

Support Vector Machines (SVM) were introduced in the early 90's as a novel nonlinear solution for classification and regression tasks. These techniques have been proved to have superior performances in a large variety of real world applications due to their generalization abilities and robustness against noise and interferences.

This book introduces a set of novel techniques based on SVM that are applied to antenna array processing and electromagnetics. In particular, it introduces methods for linear and nonlinear beamforming and parameter design for arrays and electromagnetic applications.

KEYWORDS

Support Vector Machines, Beamforming, Angle of Arrival, Electromagnetics, Antenna Arrays

Contents

Preface

This book is intended to serve as an introduction to machine learning and support vector machines for students and practicing engineers in the field of electromagnetics. The idea is to show how this combination of optimization theory, statistical learning, and kernel theory, labeled as "Support Vector Machines" can be applied to electromagnetic problems. We believe that there are still a lot of problems in electromagnetics that can be optimized using the "machine learning approach".

The basic material on support vector machines, Classifiers (SVCs) and Regressors (SVRs) are introduced in the first two chapters and then we slowly add to the their algorithmic complexity as we progress in terms of chapters. MATLAB files are provided so the readers can experiment on their own using our algorithms before they start developing their own.

An important topic that was included in this textbook is the subject of nonlinear support vector machines. This is an important subject since it addresses solution to a number of optimization problems that cannot be tackled using linear support vector machines only. The differences between linear and nonlinear SVM algorithms are discussed and their applications in electromagnetic problems are presented. The material presented here is more than enough to provide the reader of this text with some serious expertise in the subject of Support Vector Machines from the computational point of view. In fact, although this text was written with the electromagnetics engineers in mind, it can be easily used by colleagues in the fields of communications, radar and image processing. The algorithms are applicable to several optimization problems and the MATLAB files provided can be altered to fit the data at hand. For access to the electronic files of the Matlab code, please go to http://www.morganclaypool.com/page/matlab.

We hope that the reader finds the book enjoyable and helpful in triggering further research in this area.

CHAPTER 1

Introduction

1.1 MOTIVATION OF THIS BOOK

Since the 1990s there has been a significant activity in the theoretical development and applications of Support Vector Machines (SVMs) (see, e.g. [1–5]). The theory of SVMs is based on the cross-pollination of optimization theory, statistical learning, kernel theory and algorithmics. So far, machine learning has largely been devoted to solving problems relating to data mining, text categorization [6], biomedical problems as microarray data analysis [7], Magnetic Resonance Imaging [8, 9], linear signal processing [10], speech recognition [11, 12] and image processing [13–18], for example (see also [19]), but not so much in the field of electromagnetics. Recently, however, popular binary machine learning algorithms, including support vector machines (SVM), have successfully been applied to wireless communication problems [20], notably spread spectrum receiver design [21–23], and channel equalization [24]. The high-speed capabilities and "learning" abilities of support vectors can also be applied to solving complex optimization problems in electromagnetics in the areas of radar, remote sensing, microwaves and antennas.

The aim of this book is to gently introduce the subject of support vector machines in its linear and nonlinear form, both as regressors and as classifiers and to show how it can be applied to several antenna array processing problems and electromagnetics in general.

There are basically four main situations in which SVMs make good candidates for use in electromagnetics:

- When closed form solutions do not exist and trial and error methods are the only approaches to solving the problem at hand. By training an SVM one can use it to predict solutions to the problem.

- When the application requires real-time performance. This is an important feature because SVMs can be trained off-line and then implemented in hardware and embedded on any device.

- When faster convergence rates and smaller errors are required in the optimization of large systems. The superior generalized capabilities of SVM to optimize several problems make them very advantageous.

- When enough measured data exist to train an SVM for prediction purposes, especially when no analytical tools exist. In this case, one can actually use SVMs to solve part of the problem where no analytical solution exist and combine the solution with other existing analytical and closed form solutions.

1.2   LEARNING MACHINES AND GENERALIZATION

A learning machine can be formulated as a function $f(\mathbf{x})$ which makes a correspondence of the form $f : \mathbb{R}^N \rightarrow y : \mathbb{R}$ from an input sample $\mathbf{x}$ to a scalar y. Let us assume that this function is one of a class or family of parametric functions $f(\boldsymbol{w}, \mathbf{x})$, and we want to estimate the parameters $\boldsymbol{w}$ to produce the best possible estimation of the scalar y from the vector $\mathbf{x}$.

Let us assume that the pairs $\{\mathbf{x}, y\}$ are given a distribution of probability $P(\mathbf{x}, y)$, and that we provide a given loss function $\ell(f(\boldsymbol{w}, \mathbf{x}), y)$ between the estimation $f(\boldsymbol{w}, \mathbf{x})$ and the desired output y. A common loss function used in practice is the squared loss. The best possible solution would be the one that minimizes a given expected *error risk*

$$R(f(\boldsymbol{w})) = \int \ell(f(\boldsymbol{w}, \mathbf{x}), y) d P(\mathbf{x}, y) \qquad (1.1)$$

Obviously, the distribution of probability is not given in practice, so all we can do is to measure an *empirical risk* which can be expressed as the mean of the losses computed over all the available training pairs $\{\mathbf{x}, y\}$. This is often called the Empirical Risk Minimization Principle (ERM). It is possible to make this principle consistent as the number of samples goes to infinity. But for a small sample set, the difference between the actual error risk and the empirical risk may be arbitrarily large.

When using the empirical risk into the learning procedure, it learns from the data available for training. As the actual risk cannot be measured, the operational performance measure for the trained network is the error on future data outside the training set, also known as *generalization* error. This error may be undesirably large when, for example, the available training set size is too small in comparison with the network parameter set size. Practice has shown that direct minimization of the training error for a given fixed training set obtained by learning algorithms does not necessarily imply a corresponding minimization of the generalization error. In the neural network literature this phenomenon is usually referred to as *overfitting* [25].

An intuitive way to reduce overfitting is to minimize the complexity of our learning machine . The rule is: *choose the simplest possible function that satisfactorily explains the data.* This is simply the classical Occam's razor rule [1, 2]. Then, the optimization of our learning (classification or regression, for example) machine implies the selection of the model or, in other words, the measure and minimization of the complexity of this machine. If we choose a

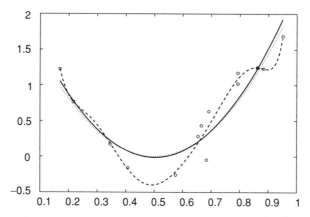

FIGURE 1.1: Example of regression using control of complexity (solid line) and without control of complexity (dashed line). The first line generalizes better, as it approaches to the theoretical function (dotted line), while the second approaches to the data available for the regression, but not to the theoretical function

complex machine as classifier or regressor (say, for example, a high degree polynomial function), this function will be able to adapt itself to the training data so it arbitrarily reduces the empirical error. If the data used for training is not representative of the statistics of the problem, then, the solution will be a poor one. This is, when using new, never seen, data to test the performance of the machine, the error may be much higher (see Fig. 1.1). A solution to reduce the test or generalization error will be to reduce the complexity of the machine, so it has less degrees of freedom to express a solution. This may be viewed as a smoothing of the learned function. The figure shows two different algorithms applied to the regression of a quadratic function. The data available for the regression operation (circles in the figure) have been corrupted with white noise. Dotted line represents the theoretical function. The solid line represents a regression in which the complexity has been controlled, while the dash line corresponds to a regression in which no control of the complexity has been applied. This line shows less quadratic error with respect to the given data, but it has a poor generalization ability, as it does not approach to the theoretical line. The solid line has a worse quadratic error with respect to the data, but its generalization ability is better.

This equivalent problem of reducing the complexity of the learning machine is commonly solved by introducing a regularization term in the function to be optimized. This term is intended to penalize the complexity of the machine [26–28]. The particular way of introducing the complexity penalty that is given by the Vapnik–Chervonenkis Theory [1,2] is the one that gives to the Support Vector Machines. In this book, we summarize the SVM from an intuitive way and then we introduce its applications to array processing.

1.3   ORGANIZATION OF THIS BOOK

This book is divided in three main parts. After the introduction, the first three chapters cover the theory of SVMS, both as classifiers and regressors. The next three chapters deal with applications in antenna array processing and other areas in electromagnetics. The four appendices at the end of the book comprise the last part. We tried to be as complete as possible with examples and the inclusion of MATLAB files that can help the readers start their application of the algorithms covered in this book.

More specifically, the 2nd Chapter starts with the introduction of linear SVM classifiers. We thought that this would be the easiest and most intuitive way to familiarize the reader with the new theoretical concepts associated with SVMs. Next, the main ideas of SVM regressors and their practical optimization are covered. The 3rd Chapter concentrates on the theory of nonlinear SVMs. The Kernel trick is formulated and the Mercer theorem introduced. Examples on how to construct nonlinear support vector classifiers and regressors are shown and MATLAB files included. The 4th Chapter covers the most recent topic of SVMs in the complex plane. Here, the foundation of SVM-autoregressor models and the concept of robust cost functions are introduced. Several ideas for the practical use of these models for optimization are presented and discussed.

The applications part of the book starts with Chapter 5. In this chapter the problem for antenna beamforming and on how to maximize the received signal-to-noise ratio are presented. This chapter really gives the reader a state of awareness on all the intricacies and difficulties associated with applying SVMs to complex problems. It is shown how to generate training and testing data that are used to optimize the bit error performance and provide a robust SVM estimator. It should also be noted here that both linear and nonlinear beamformers are constructed and compared. In Chapter 6, linear and nonlinear SVMs are applied to the angle of arrival problem (AOA). This chapter not only includes our own work but also the most recent work from other researchers in this field to demonstrate the superior optimization capabilities of SVMs. Another important topic covered in this chapter is that of the multiclass classifier. It is shown how M distinct classes one can construct $M(M-1)/2$ hyperplanes that can separate these classes with a maximum margin during the training phase of SVMs. Chapter 7 covers three more but vary different applications in the area of electromagnetics

The third and final part include several appendices with several derivations and theorems that will become very helpful to the reader who wants to develop his/her own SVM algorithms.

CHAPTER 2

Linear Support Vector Machines

We start this chapter by explaining the basic concepts of the Support Vector Classifier (SVC) [29] and the Support Vector Regressor (SVR) [30] in their linear versions. We introduce some basic examples of classification and regression that can be easily reproduced by the reader with the provided code. Our aim is to make the reader understand the underlying ideas about the SVM.

2.1 AN INTUITIVE EXPLANATION OF THE SUPPORT VECTOR CLASSIFIER

2.1.1 The Linear Classifier

Let us assume that two classes of points x_i are given in a vectorial space, and that these two classes are located in different areas in the space, so they can be separated or classified using a separating hyperplane. Let us further assume that the data is located in such a way that the separating hyperplane contains the origin (see Fig. 2.1). The expression of the separating hyperplane is then $w^T x = 0$. If one wants to know what is the class of a point x_1, one has to determine if the angle between the vector w associated with the plane and the sample x_1 is greater or less than 90°. If it is less than 90°, the sample is classified as belonging to the class A. Otherwise, the classification will be B. To do that, one computes the dot product between the vectors. If it is positive, this means that the angle is less than 90°, then the class is A. For a vector x_2, the angle will be greater than 90°, so the dot product will be negative and the class will be B. A sample located on the plane is orthogonal to w so the dot product will be zero, meaning that we cannot determine its class.

If the data is not centered around the origin, the hyperplane must be biased from it, so its expression is given by $w^T x + b = 0$.

2.1.2 A Classical Procedure to Adjust the Separating Hyperplane

The main question now is how to locate the separating hyperplane to optimally classify the data. What we really want is to minimize the probability of misclassification when classifying a set of vectors that are different from those used to adjust the parameters w and b of the hyperplane.

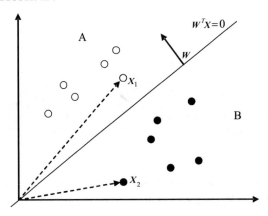

FIGURE 2.1: Classifier in a vector space

That concept is what we call *generalization ability*. Nevertheless, misclassification is measured as a binary quantity (0 for correct classification, 1 for misclassification). Instead of using these direct criteria, since the function is not differentiable, we use approximate criteria.

Classical approaches apply a procedure based on the Minimum Mean Squares Error (MMSE) between the desired result (+1 for samples of class A and −1 for samples of class B) and the actual result. The optimal separating hyperplane is the one that minimizes the mean square error between the desired and actual results obtained when classifying the given data. This criterion turns out to be optimal when the statistical properties of the data are Gaussian. But if the data is not Gaussian, the result will be biased. An easy example is shown in Fig. 2.2. In the left side, two Gaussian clusters of data are separated by a hyperplane. It is adjusted using a MMSE criterium. Samples of both classes have the minimum possible mean squared distance to the hyperplanes $\boldsymbol{w}^T\boldsymbol{x} + b = \pm 1$. But in the right side, the same procedure is applied for a data set that in some points are far from the center of the cluster, thus biasing the result.

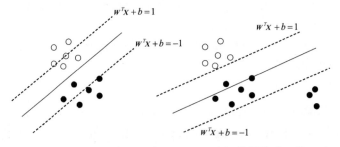

FIGURE 2.2: Left: Result of a hyperplane adjustment using MSE for Gaussian data. Right: Bias produced by the presence of outliers

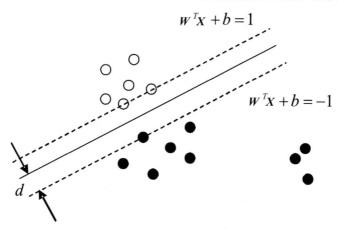

FIGURE 2.3: The SVM procedure consists of placing the hyperplane as far as possible from the nearest samples

This example may occur even in the case when the data are simply two Gaussians with outliers (samples that are statistically nonsignificant, and are far from the center of the cluster). If this occurs, the statistical properties of the data set will be biased from the theoretical ones.

2.1.3   The SVM Approach

Another procedure can be implemented which is able to deal with these situations, always with the assumption that the data are separable without misclassifications by a linear hyperplane. The optimality criterion will be: *put the hyperplane as far as possible from the nearest samples, but keeping all samples in their correct side*. This is shown in Fig. 2.3.

This translates in maximizing the margin d (see Fig. 2.3) between the separating hyperplane and its nearest samples, but now placing the margin hyperplanes $\boldsymbol{w}^T\boldsymbol{x} + b = \pm 1$ into the separation margin. One can reformulate the SVM criterion as: *maximize the distance d between the separating hyperplane and the nearest samples subject to the constraints*

$$y_i[\boldsymbol{w}^T\boldsymbol{x}_i + b] \geq 1 \qquad (2.1)$$

where $y_i \in \{+1, -1\}$ is the label associated to the sample $\boldsymbol{x}_i$.

It is straightforward to show that the distance d between the separating and the margin hyperplanes is $d = \frac{1}{||\boldsymbol{w}||^2}$. Thus, maximizing the margin d is equivalent to minimizing the norm of the vector $\boldsymbol{w}$. This provides with a more useful expression of the SVM criterion: minimize

$$L_p = ||\boldsymbol{w}||^2 \qquad (2.2)$$

subject to

$$y_i[\boldsymbol{w}^T\boldsymbol{x}_i + b] \geq 1 \qquad (2.3)$$

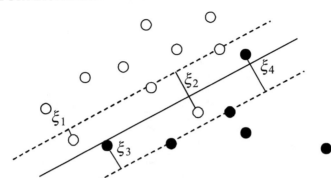

FIGURE 2.4: SVM for the nonseparable case

In practical situations, the samples are not linearly separable, so constraint (2.3) cannot be satisfied. For that reason, *slack* variables must be introduced to account for the nonseparable samples. Then, the optimization criterium consist of minimizing the (primal) functional [29]

$$L_p = \frac{1}{2}||\boldsymbol{w}||^2 + C \sum_{i=1}^{N} \xi_i \qquad (2.4)$$

subject to the constraints

$$y_i(\boldsymbol{w}^T \boldsymbol{x}_i + b) \geq 1 - \xi_i \qquad (2.5)$$

The values of ξ_i must be constrained to be nonnegative. If the sample $\boldsymbol{x}_i$ is correctly classified by the hyperplane, and it is out of the margin, its corresponding slack variable is $\xi_i = 0$. If it is well classified but it is into the margin, $0 < \xi_i < 1$. If the sample is misclassified, then $\xi_i > 1$ (Fig. 2.4). The value of C is a trade-off between the maximization of the margin and the minimization of the errors.

2.1.4   Practical Optimization of the Classifier

Once the criterion of optimality has been established, we need a method for finding the parameter vector $\boldsymbol{w}$ which meets it. The optimization problem in Eqs. *(2.4)*, *(2.5)* is a classical constrained optimization problem.

In order to solve this a optimization problem, one must apply a Lagrange optimization procedure with as many Lagrange multipliers α_i as constraints [31]. The optimization leads to the following (dual) unconstrained problem (see Appendix B): maximize

$$L_d = -\frac{1}{2} \sum_{i=1}^{N} \sum_{j=1}^{N} \alpha_i \alpha_j y_i y_j \boldsymbol{x}_i^T \boldsymbol{x}_j + \sum_{i=1}^{N} \alpha_i \qquad (2.6)$$

with $0 \leq \alpha_i \leq C$. Also, the following condition must be satisfied

$$\sum_i \alpha_i y_i = 0 \qquad (2.7)$$

Expression *(2.6)* can be rewritten into matrix format as

$$L_d = -\frac{1}{2} \boldsymbol{\alpha}^T \boldsymbol{YRY}\boldsymbol{\alpha} + \boldsymbol{\alpha} \qquad (2.8)$$

with $\boldsymbol{\alpha}$ being a column vector containing all the Lagrange multipliers α_i, $\boldsymbol{Y}$ being a diagonal matrix of the form $\boldsymbol{Y}_{ii} = y_i$, and $\boldsymbol{R}$ being the matrix of dot products

$$\boldsymbol{R}_{ij} = \boldsymbol{x}_i^T \boldsymbol{x}_j \qquad (2.9)$$

This is a quadratic functional and it can be solved using quadratic programming [32]. The solution for the parameters is given by

$$\boldsymbol{w} = \sum_{i=1}^{N} \alpha_i y_i \boldsymbol{x}_i \qquad (2.10)$$

The so-called Karush–Kuhn–Tucker conditions state that if $y_i \boldsymbol{w}^T \boldsymbol{x}_i \geq 1$ (and then $\xi_i = 0$), then $\alpha_i = 0$. In other words, only a subset of the Lagrange multipliers will have a nonzero value in the solution, while the others will vanish. Thus, the solution will be sparse and their associated samples (those for which $\alpha_i = 0$) are the so-called *Support Vectors*.

In order to compute the value of b, one has to consider that if $\alpha_i < C$, then $\xi_i = 0$, which leads to the condition

$$y_i(\boldsymbol{w}^T \boldsymbol{x}_i + b) - 1 = 0 \qquad (2.11)$$

for any sample $\boldsymbol{x}_i$ for which $\alpha_i < C$. We justify this fact in Appendix B. In practice, it is numerically convenient to average the result of b for all samples with $0 < \alpha_i < C$.

2.1.5   Example: Linear SVC in MATLAB®

A straightforward way to program a linear SVC is using the MATLAB function for quadratic programming quadprog.m. First, we generate a two-dimensional data set with some samples of two classes with this simple script:

```
x=[randn(1,10)-1  randn(1,10)+1;randn(1,10)-1 randn(1,10)+1]';
y=[-ones(1,10) ones(1,10)]';
```

This generates a matrix of 20 two-dimensional row vectors as shown in Fig. 2.5. We generate a nonseparable set. The first 10 samples are labeled as vectors of class +1, and the rest as vectors of class −1.

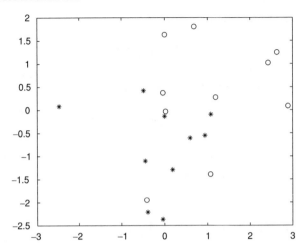

FIGURE 2.5: Generated data for the example of 2.1.5

Now, we need to solve the optimization problem in *(2.8)* for the given data using quadprog.m.

Function z=quadprog(H,f,[],[],a,K,Kl,Ku) solves the problem

$$\min_{z} \frac{1}{2}z^T H z + f^T z$$

subject to

$$az = K$$
$$K_l \leq z \leq K_u$$

In order to adapt the function to the SVC optimization in *(2.8)*, we just need to identify the variables $z = \alpha$, $H = YRY$, $f = -1$, $a = y^T$, $K = 0$, $K_l = 0$ and $K_u = C1$. Matrix H may be ill-conditioned [27], i.e., it may have some very small eigenvalues. In order to avoid numerical inaccuracies, the matrix must be numerically regularized by adding a small identity matrix. We discuss the mean of this numerical regularization in Section 4.3.

The MATLAB code for the optimization will be, then

```
R=x*x';                             %Dot products
Y=diag(y);
H=Y*R*Y+1e-6*eye(length(y));        %Matrix H regularized
f=-ones(size(y)); a=y'; K=0;
Kl=zeros(size(y));
C=100;                              %Functional Trade-off
Ku=C*ones(size(y));
```

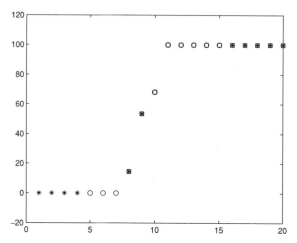

FIGURE 2.6: Values of the Lagrange multipliers α_i. Circles and stars correspond to the circles and squares of the data in Figs. 2.5 and 2.7

```
alpha=quadprog(H,f,[],[],a,K,Kl,Ku); %Solver
w=x'*(alpha.*y);                      %Parameters of the
                                      %Hyperplane
```

The values of the Lagrange multipliers are shown in Fig. 2.6. It can be seen that most of them have vanished. Some are saturated to C and three of them are between 0 and C. The last three ones correspond to the samples which are exactly in the margin.

Finally, we must find the bias b of the separating hyperplane, and we can do that from the fact that all samples $\boldsymbol{x}_i$ for which $0 < \alpha_i < C$ satisfy $y_i(\boldsymbol{w}^T\boldsymbol{x}_i) + b - 1 = 0$. We compute b for all those samples and then we average the results. The MATLAB code is

```
e=1e-6;                          %Tolerance to errors in alpha
ind=find(alpha>e & alpha<C-e)    %Search for 0 < alpha i < C
b=mean(y(ind) - x(ind,:)*w)      %Averaged result
```

A plot of the resulting separating and the margin hyperplanes can be seen in Fig. 2.7. For this example, three of the vectors are in the margin, as stated before. The vectors marked with squares are the Support Vectors, this is, the ones with nonzero Lagrange multiplier α_i.

2.1.6   Summary of the SVC MATLAB® Code

Here we summarize the code used to obtain the results of the presented example. The data has been generated sampling 2 subsets of 10 two-dimensional data points. Each subset has a Gauss

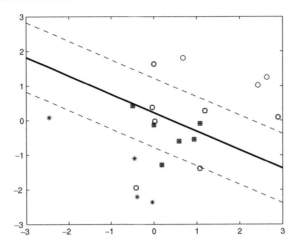

FIGURE 2.7: Resulting margin and separating hyperplanes. Support Vectors are marked

distribution with mean $(-1,1)$ and $(1,1)$. The first subset has been labeled with the label "-1" and the second subset has been labeled with "1".

The optimization has been performed using the MATLAB function `quadprog`. The representation of the solution together with the samples used to train the classifier is also detailed.

```
%%%%%%%%%%%%%%%%%%%%%
%%% Data Generation %%%
%%%%%%%%%%%%%%%%%%%%%%%%%
x=[randn(1,10)-1  randn(1,10)+1;randn(1,10)-1 randn(1,10)+1]';
y=[-ones(1,10) ones(1,10)]';



%%%%%%%%%%%%%%%%%%%%%%%%%%
%%% SVC Optimization %%%
%%%%%%%%%%%%%%%%%%%%%%%%%%
R=x*x';                          % Dot products
Y=diag(y);
H=Y*R*Y+1e-6*eye(length(y));     % Matrix H regularized
f=-ones(size(y)); a=y'; K=0;
Kl=zeros(size(y));
C=100;                           % Functional Trade-off
Ku=C*ones(size(y));
alpha=quadprog(H,f,[],[],a,K,Kl,Ku); % Solver
w=x'*(alpha.*y);                 % Parameters of the Hyperplane
```

```
%%% Computation of the bias b %%%
e=1e-6;                          % Tolerance to errors in alpha
ind=find(alpha>e & alpha<C-e)    % Search for 0 < alpha_i < C
b=mean(y(ind) - x(ind,:)*w)      % Averaged result
```

For the sake of brevity, we include below just a simplified code of representation of the data for the reader to use it.

```
%%%%%%%%%%%%%%%%%%%%%%%%
%%% Representation %%%
%%%%%%%%%%%%%%%%%%%%%%%%
data1=x(find(y==1),:);
data2=x(find(y==-1),:);
svc=x(find(alpha>e),:);

plot(data1(:,1),data1(:,2),'o')
hold on
plot(data2(:,1),data2(:,2),'*')
plot(svc(:,1),svc(:,2),'s')

% Separating hyperplane
plot([-3 3],[(3*w(1)-b)/w(2) (-3*w(1)-b)/w(2)])
% Margin hyperplanes
plot([-3 3],[(3*w(1)-b)/w(2)+1 (-3*w(1)-b)/w(2)+1],'--')
plot([-3 3],[(3*w(1)-b)/w(2)-1 (-3*w(1)-b)/w(2)-1],'--')
```

Finally, the reader may generate a test set using the same code used for the training data generation, and a test may be performed by adding the code lines

```
%%%%%%%%%%%%%%%%%%%%%%%%%%%%%%%%
%%% Test Data Generation %%%
%%%%%%%%%%%%%%%%%%%%%%%%%%%%%%%%
x=[randn(1,10)-1  randn(1,10)+1;randn(1,10)-1 randn(1,10)+1]';
y=[-ones(1,10) ones(1,10)]';

y_pred=sign(x*w+b);     %Test
error=mean(y_pred~=y);  %Error Computation
```

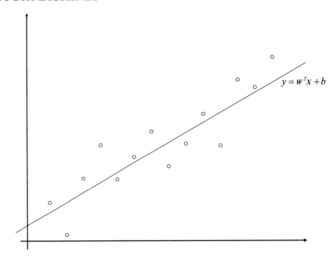

FIGURE 2.8: Linear regressor

2.2   AN INTUITIVE EXPLANATION OF THE SUPPORT VECTOR REGRESSOR

2.2.1   The Main Idea

A linear regressor is a function $f(x) = w^T x + b$ which approximates a given mapping from a set of vectors $x \in \mathbb{R}^n$ to a set of scalars $y \in \mathbb{R}$ (see Fig. 2.8).

As in classification, linear regression has been traditionally solved using Least Squares approaches. In other words, the regressor function is the hyperplane that fits the given data with the minimum mean square error between itself and the data.

The Support Vector Regressor (SVR) has a different objective. The main idea of a SVR is to find a function which fits the data with a deviation less than a given quantity ε for every single pair x_i, y_i. At the same time, we want the solution to have a minimum norm $\|w\|$. This means that SVR does not minimize errors less than ε, but only higher errors. Doing this allows us to construct a machine whose parameters are a linear combination of those samples whose error is higher or equal than ε. This leads to a sparse solution, as in SVC.

Also, minimizing the norm produces a smooth solution, or a solution which minimizes overfitting. The idea is *to find a solution which uses the minimum possible energy of the data.*

2.2.2   Formulation of the SVR

The above idea to adjust the linear regressor can be formulated in the following primal functional, in which we minimize the total error plus the norm of w.

$$L_p = \frac{1}{2}\|w\|^2 + C \sum_{i=1}^{N}(\xi_i + \xi_i')$$

(2.12)

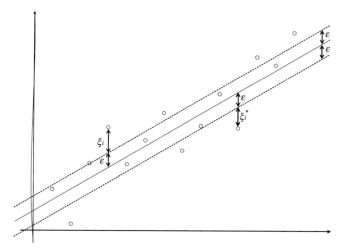

FIGURE 2.9: Concept of ε-insensitivity. Only the samples out of the $\pm\varepsilon$ margin will have a nonzero slack variable, so they will be the only ones that will be part of the solution

subject to the constraints

$$y_i - \boldsymbol{w}^T\boldsymbol{x}_i - b \leq \xi_i + \varepsilon$$
$$-y_i + \boldsymbol{w}^T\boldsymbol{x}_i + b \leq \xi_i' + \varepsilon \qquad (2.13)$$
$$\xi_i, xi_i' \geq 0$$

Constraints *(2.13)* have the following meaning: for each sample, if the error is positive (negative) and its absolute value is higher than ε, then this absolute value is forced to be less than $\xi_i + \varepsilon$ $(\xi_i' + \varepsilon)$. If the absolute value of the error is less than ε, then the corresponding slack variable will be zero, as this is the minimum allowed value for the slack variables in the constraints *(2.13)*. This is the concept of ε-*insensitivity* [2]. The concept is illustrated in Fig. 2.9.

The functional is intended to minimize the sum of the slack variables ξ_i, xi_i'. Only losses of samples for which the error is greater than ε appear, so the solution will be only function of those samples. The applied cost function is a linear one, so the described procedure is equivalent to the application of the so-called Vapnik or ε-insensitive cost function (shown in Fig. 2.10)

$$\ell(e_i) = \begin{cases} 0 & |e_i| < \varepsilon \\ |e_i| - \varepsilon & |e_i| > \varepsilon \end{cases} \qquad (2.14)$$

for $e = \xi_i + \varepsilon, e = -\xi_i' - \varepsilon$.

This procedure is similar to the one applied to SVC. In principle, we should force the errors to be less than ε and we minimize the norm of the parameters. Nevertheless, in practical situations, it may be not possible to force all the errors to be less than ε. In order to be able

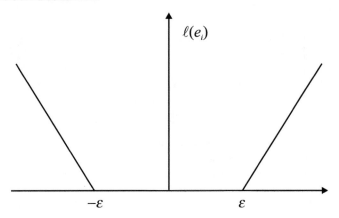

FIGURE 2.10: Vapnik or ε-insensitive cost function

to solve the functional, we introduce slack variables on the constraints, and then we minimize them.

2.2.3   Practical Optimization of the SVR

In order to solve the constrained optimization problem in *(2.12)*, *(2.13)*, we must apply Lagrange optimization to convert it into an unconstrained one. The resulting dual formulation for this dual functional is

$$L_d = -\frac{1}{2}\sum_{i=1}^{N}\sum_{i=1}^{N}(\alpha_i - \alpha_i')x_i^T x_j(\alpha_i - \alpha_j')$$
$$+ \sum_{i=1}^{N}((\alpha_i - \alpha_i')y_i - (\alpha_i + \alpha_i')\varepsilon)$$

(2.15)

with the additional constraint

$$0 \leq (\alpha_i - \alpha_i') \leq C$$

(2.16)

The important result of this derivation is that the expression of the parameters $\boldsymbol{w}$ is

$$\boldsymbol{w} = \sum_{i=1}^{N}(\alpha_i - \alpha_i')x_i$$

(2.17)

and that

$$\sum_{i=1}^{N}(\alpha_i - \alpha_i') = 0$$

(2.18)

In order to find the bias, we just need to recall that for all samples that lie in one of the two margins, the error is exactly ε for those samples $\alpha_i, \alpha_i' < C$ (see Appendix C for details).

Once these samples are identified, we can solve b from the following equations

$$-y_i + \boldsymbol{w}^T \boldsymbol{x}_i + b + \varepsilon = 0$$
$$y_i - \boldsymbol{w}^T \boldsymbol{x}_i - b + \varepsilon = 0 \qquad (2.19)$$
$$\text{for those } x_i \text{ for which } \alpha_i, \alpha_i' < C$$

A complete derivation of this functional can also be found in Appendix C.

In matrix notation we get

$$L_d = -\frac{1}{2}(\boldsymbol{\alpha} - \boldsymbol{\alpha}')^T \boldsymbol{R}(\boldsymbol{\alpha} - \boldsymbol{\alpha}') + (\boldsymbol{\alpha} - \boldsymbol{\alpha}')^T \boldsymbol{y} - (\boldsymbol{\alpha} + \boldsymbol{\alpha}')\mathbf{1}\varepsilon \qquad (2.20)$$

This functional can be maximized using the same procedure used for SVC. Very small eigenvalues may eventually appear in the matrix, so it is convenient to numerically regularize it by adding a small diagonal matrix to it. The functional becomes

$$L_d = -\frac{1}{2}(\boldsymbol{\alpha} - \boldsymbol{\alpha}')^T [\boldsymbol{R} + \gamma \boldsymbol{I}](\boldsymbol{\alpha} - \boldsymbol{\alpha}') + (\boldsymbol{\alpha} - \boldsymbol{\alpha}')^T \boldsymbol{y} - (\boldsymbol{\alpha} + \boldsymbol{\alpha}')\mathbf{1}\varepsilon \qquad (2.21)$$

In Section 4.3 it is shown that this numerical regularization is equivalent to the application of a modified version of the cost function. The interpretation of the regularization constant γ in terms of this cost function is derived.

2.2.4 Example: Linear SVR in MATLAB®

We start by writing a simple linear model of the form

$$y(x_i) = ax_i + b + n_i \qquad (2.22)$$

where x_i is a random variable and n_i is a Gaussian process. The code is

```
x=rand(100,1)              % Generates 100 uniform
                           % samples between -1 and 1
y=1.5*x+1+0.1*randn(100,1) % Linear model plus noise
```

A sample of the generated data is shown in Fig. 2.11.

We need to compute the dot product matrix $\boldsymbol{R}$ and then the product $(\boldsymbol{\alpha} - \boldsymbol{\alpha}')^T [\boldsymbol{R} + \gamma \boldsymbol{I}](\boldsymbol{\alpha} - \boldsymbol{\alpha}')$, but the Lagrange multipliers α_i and α_i' should be split to be identified after the optimization. To achieve this we use the equivalent form

$$\begin{pmatrix} \boldsymbol{\alpha} \\ \boldsymbol{\alpha}' \end{pmatrix}^T \left[\begin{pmatrix} \boldsymbol{R} & -\boldsymbol{R} \\ -\boldsymbol{R} & \boldsymbol{R} \end{pmatrix} + \gamma \begin{pmatrix} \boldsymbol{I} & -\boldsymbol{I} \\ -\boldsymbol{I} & \boldsymbol{I} \end{pmatrix} \right] \begin{pmatrix} \boldsymbol{\alpha} \\ \boldsymbol{\alpha}' \end{pmatrix} \qquad (2.23)$$

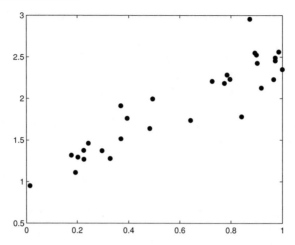

FIGURE 2.11: Generated data for the SVR example

The code for the regularized matrix is

```
R_=x*x';
R_=R_+1e-9*eye(size(R_));
H=[R_ -R_;-R_ R_];
```

We make use of the function `quadprog.m`, now to solve the optimization problem in *(2.21)*. Remind that function

```
z=quadprog(H,f,[],[],a,K,Kl,Ku)
```

solves the problem

$$\min_{z} \frac{1}{2}z^T H z + f^T z$$

subject to

$$az = K$$
$$Kl \leq z \leq Ku$$

and we are solving the optimization problem

$$L_d = -\frac{1}{2}(\alpha - \alpha')^T[R + \gamma I](\alpha - \alpha') + (\alpha - \alpha')^T y - (\alpha + \alpha')1\varepsilon$$

with the constraints

$$0 < \alpha_i, \alpha_i' < C$$
$$\sum_{i=1}^{N}(\alpha_i - \alpha_i') = 0$$

so we simply need to identify $az = K$ with $\binom{\alpha}{\alpha'}^{T}\binom{1}{-1'} = 0$ from condition *(2.18)*. Also, we identify $Kl \leq z \leq Ku$ with $0 \leq \binom{\alpha}{\alpha'} \leq 1C$ from condition *(2.16)* and, finally, $f = \binom{y}{-y} - 1\varepsilon$. Then, the code for the optimization is

```
a=[ones(size(y')) -ones(size(y'))];
y2=[y;-y];
epsilon=0.1; C=100;
f=-y2'+epsilon*ones(size(y2'));
K=0;
K1=zeros(size(y2'));
Ku=C*ones(size(y2'));
alpha=quadprog(H,f,[],[],a,K,K1,Ku);      % Solver
```

Using expression *(2.17)*, we find the coefficients of the regressor

```
beta=(alpha(1:end/2)-alpha(end/2+1:end));
w=beta'*x;
```

Finally, we solve the bias using the expression *(2.19)* and averaging

```
e=1e-6;                              % Tolerance to errors
                                     % in alpha
ind=find(abs(beta)>e & abs(beta)<C-e)  % Search for
                                     % 0 < alpha_i < C
b=mean(y(ind) - x(ind,:)*w)          % Averaged result
```

A plot of the result is depicted in Fig. 2.12. The continuous line represents the resulting SVR line, and the dotted line is the actual linear model. Dashed lines represent the margins and the squared points are the Support Vectors. Note that all support vectors are out or on the margin, but not into it.

2.2.5   Summary of the SVR MATLAB® Code

```
%%%%%%%%%%%%%%%%%%%%%%%%
%%% Data Generation %%%
%%%%%%%%%%%%%%%%%%%%%%%%
x=rand(30,1);                        % Generate 30 samples
y=1.5*x+1+0.2*randn(30,1);           % Linear model plus noise
```

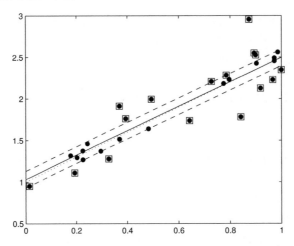

FIGURE 2.12: Result of the SVR procedure. Continuous line: SVR; Dotted line: actual linear model; Dashed lines: margins; Squared points: Support Vectors

```
%%%%%%%%%%%%%%%%%%%%%%
%%% SVR Optimization %%%
%%%%%%%%%%%%%%%%%%%%%%%%%
R_=x*x';
R=[R_ -R_;-R_ R_];
a=[ones(size(y')) -ones(size(y'))];
y2=[y;-y];
H=(R+1e-9*eye(size(R,1)));
epsilon=0.1; C=100;
f=-y2'+epsilon*ones(size(y2'));
K=0;
K1=zeros(size(y2'));
Ku=C*ones(size(y2'));
alpha=quadprog(H,f,[],[],a,K,K1,Ku);      % Solver

beta=(alpha(1:end/2)-alpha(end/2+1:end));
w=beta'*x;
%% Computation of bias b %%
e=1e-6;                                    % Tolerance to errors in
                                           %   alpha
ind=find(abs(beta)>e & abs(beta)<C-e)  % Search for
                                       % 0 < alpha_i < C
b=mean(y(ind) - x(ind,:)*w)            % Averaged result
```

```
%%%%%%%%%%%%%%%%%%%%%
%%% Representation %%%
%%%%%%%%%%%%%%%%%%%%%%
plot(x,y,'.')                              % All data
hold on
ind=find(abs(beta)>e);
plot(x(ind),y(ind),'s')                    % Support Vectors
plot([0 1],[b w+b])                        % Regression line
plot([0 1],[b+epsilon w+b+epsilon],'--') % Margins
plot([0 1],[b-epsilon w+b-epsilon],'--')
plot([0 1],[1 1.5+1],':')                  % True model}
```

CHAPTER 3

Nonlinear Support Vector Machines

3.1 THE KERNEL TRICK

The practical application of the Support Vector Machine procedure is only possible with linear machines because we only have an optimality criterion for linear hyperplanes. There are no general results for nonlinear functions. Nevertheless, for many classification or regression problems, a linear solution does not provide good performances, so many times a nonlinear approach is needed.

Fortunately, there is a theorem provided by Mercer (see, e.g. [33, 34]) in the early 1900s which is of extreme relevance to extend the principle to nonlinear machines. The basic idea is that vectors x in a finite dimension space (called input space) can be mapped to a higher (possibly infinite) dimensional Hilbert[1] space $\mathcal{H}$ provided with a dot product through a nonlinear transformation $\varphi(\cdot)$. A linear machine can be constructed in a higher dimensional space [1, 2] (often called the feature space), but it stays nonlinear in the input space. This is possible by virtue of the Mercer theorem.

Most of the transformations $\varphi(\cdot)$ are unknown, but the dot product of the corresponding spaces can be expressed as a function of the input vectors as

$$\varphi(\boldsymbol{x}_i)^T \varphi(\boldsymbol{x}_j) = K(\boldsymbol{x}_i, \boldsymbol{x}_j) \qquad (3.1)$$

These spaces are called Reproducing Kernel Hilbert Spaces (RKHS), and their dot products $K(\boldsymbol{x}_i, \boldsymbol{x}_j)$ are called Mercer kernels. Fortunately, an explicit representation of the vectors in the feature space is not required as the SVM formulation only contains dot products among vectors. Thus, it is possible to explicitly represent an SVM into a Hilbert space.

The Mercer theorem gives the condition that a kernel $K(\boldsymbol{x}_i, \boldsymbol{x}_j)$ must satisfy in order to be the dot product of a Hilbert space. There exist a function $\varphi : \mathcal{R}^n \rightarrow \mathcal{H}$ and a dot product

---

[1]A Hilbert space is a generalization of an Euclidean space, specifically any linear space provided with an inner product and that is complete with respect to this defined norm (i.e. any Cauchy sequence converges to a point in these space).

$K(\boldsymbol{x}_i, \boldsymbol{x}_j) = \varphi(\boldsymbol{x}_i)^T \varphi(\boldsymbol{x}_j)$ if and only if for any function $g(\boldsymbol{x})$ for which

$$\int g(\boldsymbol{x})d\boldsymbol{x} < \infty \tag{3.2}$$

the inequality

$$\int K(\boldsymbol{x}, \boldsymbol{y})g(\boldsymbol{x})g(\boldsymbol{y})d\boldsymbol{x}\,d\boldsymbol{y} \geq 0 \tag{3.3}$$

is satisfied.

This condition is not always easy to prove for any function. The first kernels to be proven to fit the Mercer theorem were the homogeneous polynomial kernel

$$K(\boldsymbol{x}_i, \boldsymbol{x}_j) = (\boldsymbol{x}_i^T \boldsymbol{x}_j)^p \tag{3.4}$$

and the inhomogeneous polynomial kernel

$$K(\boldsymbol{x}_i, \boldsymbol{x}_j) = \left(\boldsymbol{x}_i^T \boldsymbol{x}_j + 1\right)^p \tag{3.5}$$

The mapping for these two kernel families can be found in an explicit way and the corresponding Hilbert space has finite dimension equal to $\binom{d+n-1}{p}$ where n is the dimension of the input space.

Another important kernel type is the Gaussian kernel, because it is widely used in many practical applications. Its expression is

$$K(\boldsymbol{x}_i, \boldsymbol{x}_j) = e^{\frac{||\boldsymbol{x}_i - \boldsymbol{x}_j||^2}{2\sigma^2}} \tag{3.6}$$

The corresponding nonlinear mapping is not explicit and the dimension of the Hilbert space is infinite.

Other kernel examples are the sigmoidal kernel

$$K(\boldsymbol{x}_i, \boldsymbol{x}_j) = \tanh\left(\gamma \boldsymbol{x}_i^T \boldsymbol{x}_j + \mu\right) \tag{3.7}$$

and the Dirichlet kernel

$$K(\boldsymbol{x}_i, \boldsymbol{x}_j) = \frac{\sin((n + 1/2)(\boldsymbol{x}_i - \boldsymbol{x}_j))}{2 \sin((\boldsymbol{x}_i - \boldsymbol{x}_j)/2)} \tag{3.8}$$

3.2   CONSTRUCTION OF A NONLINEAR SVC

The solution for the linear Support Vector Classifier is given by a linear combination of a subset of the training data

$$\boldsymbol{w} = \sum_{i=1}^{N} y_i \alpha_i \boldsymbol{x}_i$$

If prior to the optimization the data is mapped into a Hilbert space, then the solution becomes

$$w = \sum_{i=1}^{N} y_i \alpha_i \varphi(x_i) \qquad (3.9)$$

where $\varphi(\cdot)$ is the mapping function.

The parameter vector w is a combination of vectors into the Hilbert space, but recall that many transformations $\varphi(\cdot)$ are unknown. Thus, we may not have an explicit form for them. But the problem can still be solved, because the machine just needs the dot products of the vector, and not an explicit form of them. We cannot use this

$$y_j = w^T \varphi(x_j) + b \qquad (3.10)$$

because the parameters w are in an infinite-dimensional space, so no explicit expression exists for them. However, by substituting Eq. *(3.9)* into *(3.10)*, yields

$$y_j = \sum_{i=1}^{N} y_i \alpha_i \varphi(x_i)^T \varphi(x_j) + b = \sum_{i=1}^{N} y_i \alpha_i K(x_i, x_j) + b \qquad (3.11)$$

The resulting machine can now be expressed directly in terms of the Lagrange multipliers and the Kernel dot products . In order to solve the dual functional which determines the Lagrange multipliers, the vectors are not required either, but only the Gram matrix K of the dot products between them. Again, the kernel is used to compute this matrix

$$K_{ij} = K(x_i, x_j) \qquad (3.12)$$

Once this matrix has been computed, solving for a nonlinear SVM is as easy as solving for a linear one, as long as the matrix is positive definite. It can be shown that if the kernel fits the Mercer theorem, the matrix will be positive definite [34].

In order to compute the bias b, we can still make use of the expression *(2.11)*, but for the nonlinear SVC it becomes

$$y_j \left(\sum_{i=1}^{N} y_i \alpha_i \varphi(x_i)^T \varphi(x_j) + b \right) - 1 = 0$$
$$y_j \left(\sum_{i=1}^{N} y_i \alpha_i K(x_i, x_j) + b \right) - 1 = 0 \qquad (3.13)$$

for all x_j for which $\alpha_j < C$

We just need to extract b from expression *(3.13)* and average it for all samples with $\alpha_i < C$.

3.2.1 Example: Nonlinear SVC in MATLAB®

In this example, we try to classify a set of data which cannot be reasonably classified using a linear hyperplane. We generate a set of 40 training vectors using this code

```
k=20; %Number of training data per class
ro=2*pi*rand(k,1);
r=5+randn(k,1);
x1=[r.*cos(ro) r.*sin(ro)];
x2=[randn(k,1) randn(k,1)];
x=[x1;x2];
y=[-ones(1,k) ones(1,k)]';
```

Also, we generate a set of 100 vectors to test the machine

```
ktest=50; %Number of test data per class
ro=2*pi*rand(ktest,1);
r=5+randn(ktest,1);
x1=[r.*cos(ro) r.*sin(ro)];
x2=[randn(ktest,1) randn(ktest,1)];
xtest=[x1;x2];
ytest=[-ones(1,ktest) ones(1,ktest)]';
```

A sample of the generated data is shown in Fig. 3.1. The next step for the SVC procedure is to compute the dot product matrix. Since we want a nonlinear classifier, we compute this dot product matrix using a kernel. We choose the Gaussian kernel $K(u, v) = \exp\left(\frac{-|u-v|^2}{2\sigma}\right)$. The computation of the kernel matrix of *(3.12)* implements as follows. First, we need to compute the matrix of all the distances between vectors

```
N=2*k;          % Number of data
sigma=1;        % Parameter of the kernel
D=buffer(sum([kron(x,ones(N,1))...
  - kron(ones(1,N),x')'].^2,2),N,0)
% This is a recipe for fast computation
% of a matrix of distances in MATLAB
% using the Kronecker product
```

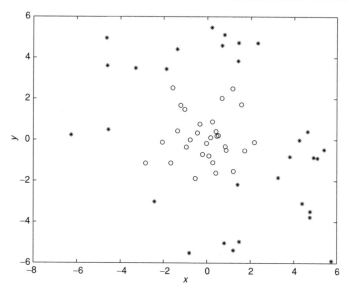

FIGURE 3.1: Data for the nonlinear SVC example

Then we compute the kernel matrix

```
R=exp(-D/(2*sigma)); % Kernel Matrix
```

Once the matrix has been obtained, the optimization procedure is exactly equal to the one for the linear case, except for the fact that we cannot have an explicit expression for the parameters w.

```
Y=diag(y);
H=Y*R*Y+1e-6*eye(length(y));          % Matrix H regularized
f=-ones(size(y)); a=y'; K=0;
Kl=zeros(size(y));
C=100;                                 % Functional Trade-off
Ku=C*ones(size(y));
alpha=quadprog(H,f,[],[],a,K,Kl,Ku);  % Solver
ind=find(alpha>e);
x_sv=x(ind,:);                         % Extraction of the
                                       % support vectors
N_SV=length(ind);                      % Number of support vectors
```

To compute b, we find first all vectors x_i with $\alpha_i < C$. We use *(3.13)* to extract b for all these vectors and finally average the results.

```
%%% Computation of the bias b %%%
e=1e-6;                              % Tolerance to errors
                                     % in alpha
ind=find(alpha>e & alpha<C-e);       % Search for
                                     % 0 < alpha_i < C
N_margin=length(ind);

D=buffer(sum([kron(x_sv,ones(N_margin,1))...
   - kron(ones(1,N_SV),x(ind,:)')')'].^2,2),N_margin,0);

% Computation of the kernel matrix
R_margin=exp(-D/(2*sigma));
y_margin=R_margin*(y(ind).*alpha(ind));
b=mean(y(ind) - y_margin);           % Averaged result
```

The next step is to construct the Support Vector Classifier using the expression *(3.11)*. We first compute the matrix of distances between the support vectors and the test vectors

```
ind=find(alpha>e);
x_sv=x(ind,:); % Extraction of the support vectors
N_SV=length(ind); % Number of support vectors
N_test=2*ktest; % Number of test data
%%Computation of the kernel matrix%%
D=buffer(sum([kron(x_sv,ones(N_test,1))...
   - kron(ones(1,N_SV),xtest')')'].^2,2),N_test,0);
```

and then we compute the kernel matrix corresponding to *(3.11)*

```
R_test=exp(-D/(2*sigma));
```

Finally, the support vector machine can be computed as

```
y_output=sign(R_test*(y(ind).*alpha(ind))+b);
```

We can approximately draw the separating boundary in the input space using numerical computations (see Subsection 3.2.2 for the details). The result is shown in Fig. 3.2.

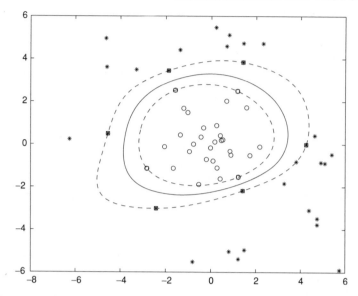

FIGURE 3.2: Separating boundary, margins and support vectors for the nonlinear SVC example

3.2.2   Summary of the Code

```
%%%%%%%%%%%%%%%%%%%%%%%
%%% Data Generation %%%
%%%%%%%%%%%%%%%%%%%%%%%
k=20;                                    %Number of training data per
                                         class

ro=2*pi*rand(k,1);
r=5+randn(k,1);
x1=[r.*cos(ro) r.*sin(ro)];
x2=[randn(k,1) randn(k,1)];
x=[x1;x2];
y=[-ones(1,k) ones(1,k)]';


ktest=50;                                %Number of test data per class
ro=2*pi*rand(ktest,1);
r=5+randn(ktest,1);
x1=[r.*cos(ro) r.*sin(ro)];
x2=[randn(ktest,1) randn(ktest,1)];
xtest=[x1;x2];
ytest=[-ones(1,ktest) ones(1,ktest)]';
```

```matlab
%%%%%%%%%%%%%%%%%%%%%%%
%%% SVC Optimization %%%
%%%%%%%%%%%%%%%%%%%%%%%%%
N=2*k;                              % Number of data
sigma=2;                            % Parameter of the kernel
D=buffer(sum([kron(x,ones(N,1))...
 - kron(ones(1,N),x')'].^2,2),N,0);
% This is a recipe for fast computation
% of a matrix of distances in MATLAB
R=exp(-D/(2*sigma));                % Kernel Matrix


Y=diag(y);
H=Y*R*Y+1e-6*eye(length(y));        % Matrix H regularized
f=-ones(size(y)); a=y'; K=0;
Kl=zeros(size(y));
C=100;                              % Functional Trade-off
Ku=C*ones(size(y));
alpha=quadprog(H,f,[],[],a,K,Kl,Ku);  % Solver
ind=find(alpha>e);
x_sv=x(ind,:);                      % Extraction of the support
                                    % vectors

N_SV=length(ind);                   % Number of SV


%%% Computation of the bias b %%%
e=1e-6;                             % Tolerance to
                                    % errors in alpha

ind=find(alpha>e & alpha<C-e);      % Search for
                                    % 0 < alpha_i < C
N_margin=length(ind);
D=buffer(sum([kron(x_sv,ones(N_margin,1)) ...
        - kron(ones(1,N_SV),x(ind,:)')'].^2,2),N_margin,0);
% Computation of the kernel matrix
R_margin=exp(-D/(2*sigma));
y_margin=R_margin*(y(ind).*alpha(ind));
b=mean(y(ind) - y_margin);          % Averaged result
```

```
%%%%%%%%%%%%%%%%%%%%%%%%%%%%%%%%%%%%
%%% Support Vector Classifier %%%
%%%%%%%%%%%%%%%%%%%%%%%%%%%%%%%%%%%%

N_test=2*ktest;                          % Number of test data
%%Computation of the kernel matrix%%
D=buffer(sum([kron(x_sv,ones(N_test,1))...
        - kron(ones(1,N_SV),xtest')'].^2,2),N_test,0);
% Computation of the kernel matrix
R_test=exp(-D/(2*sigma));
% Output of the classifier
y_output=sign(R_test*(y(ind).*alpha(ind))+b);
errors=sum(ytest~=y_output)              % Error Computation


%%%%%%%%%%%%%%%%%%%%%%%%
%%% Representation %%%
%%%%%%%%%%%%%%%%%%%%%%%%
data1=x(find(y==1),:);
data2=x(find(y==-1),:);
svc=x(find(alpha>e),:);


plot(data1(:,1),data1(:,2),'o')
hold on
plot(data2(:,1),data2(:,2),'*')
plot(svc(:,1),svc(:,2),'s')


g=(-8:0.1:8)';                           % Grid between -8 and 8
x_grid=[kron(g,ones(length(g),1)) kron(ones(length(g),1),g)];
N_grid=length(x_grid);
D=buffer(sum([kron(x_sv,ones(N_grid,1))...
  - kron(ones(1,N_SV),x_grid')'].^2,2),N_grid,0);
% Computation of the kernel matrix
R_grid=exp(-D/(2*sigma));
y_grid=(R_grid*(y(ind).*alpha(ind))+b);
contour(g,g,buffer(y_grid,length(g),0),[0 0]) % Boundary draw
```

3.3   CONSTRUCTION OF A NONLINEAR SVR

The solution for the linear Support Vector Regressor is

$$\boldsymbol{w} = \sum_{i=1}^{N} (\alpha_i - \alpha_i')\boldsymbol{x}_i$$

Its nonlinear counterpart will have the expression

$$\boldsymbol{w} = \sum_{i=1}^{N} (\alpha_i - \alpha_i')\varphi(\boldsymbol{x}_i) \tag{3.14}$$

Following the same procedure as in the SVC, one can find the expression of the nonlinear SVR

$$y_j = \sum_{i=1}^{N} (\alpha_i - \alpha_i')\varphi(\boldsymbol{x}_i)^T\varphi(\boldsymbol{x}_j) + b = \sum_{i=1}^{N} (\alpha_i - \alpha_i')K(\boldsymbol{x}_i, \boldsymbol{x}_j) + b \tag{3.15}$$

The construction of a nonlinear SVR is almost identical to the construction of the nonlinear SVC. We let the reader the coding of these machine as an exercise.

CHAPTER 4

Advanced Topics

4.1 SUPPORT VECTOR MACHINES IN THE COMPLEX PLANE

In communications applications the used signals are complex-valued. In order to process them, it is not a good idea to use two separate processors for real and imaginary parts, since there is cross-information that cannot be ignored.

So, there is a need for a complex-valued version of support vector machines. This version can be found in taking into account the in-phase or real and quadrature-phase or imaginary parts of the error signal.

We present here a summarized derivation of the complex-valued linear SVM regressor, but similar derivations hold true for other SVM approaches. Also, using the Kernel trick of Section 3.1, the result is directly applicable to nonlinear machines.

As in the real regressor, the risk function consists of two terms, a structural risk term, which comprises a bound on the complexity of the resulting model and an empirical risk term, measured directly on the residuals by using an appropriate cost function. Now the residuals are complex-valued, so we need to minimize simultaneously both real and imaginary parts of them. Therefore, applying the error cost function *(2.14)* is equivalent to minimize

$$L_p = \frac{1}{2}||\boldsymbol{w}||^2 + C \sum_{i=1}^{N}(\xi_n + \xi_n') + C \sum_{i=1}^{N}(\zeta_n + \zeta_n') \qquad (4.1)$$

subject to

$$
\begin{aligned}
Re\left(y_i - \boldsymbol{w}^T\mathbf{x}_i - b\right) &\leq \varepsilon + \xi_i \\
Re\left(-y_i + \boldsymbol{w}^T\mathbf{x}_i + b\right) &\leq \varepsilon + \xi_i' \\
Im\left(y_i - \boldsymbol{w}^T\mathbf{x}_i - b\right) &\leq \varepsilon + \zeta_i \\
Im\left(-y_i + \boldsymbol{w}^T\mathbf{x}_i + b\right) &\leq \varepsilon + \zeta_i' \\
\xi_i, \xi_i', \zeta_i, \zeta_i' &\geq 0
\end{aligned}
\qquad (4.2)
$$

where ξ_i (ξ_i') are the slack variables or losses standing for positive (negative) errors in the real part of the output -analogously for ζ_i (ζ_i') in the imaginary part (Fig. 4.1).

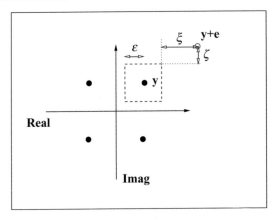

FIGURE 4.1: Complex-valued single sample y, its ε-insensitivity zone, and relationship between errors (e) and losses

The primal-dual Lagrange functional can be written with Lagrange multipliers α_n, β_n, λ_n, η_n, α'_n, β'_n, λ'_n, $\eta'_n \geq 0$.

$$
\begin{aligned}
L_{pd} = \frac{1}{2}||\boldsymbol{w}||^2 &+ C\sum_{i \in I_1}(\xi_i + \xi'_i) + C\sum_{i \in I_1}(\zeta_i + \zeta'_i) \\
&- \sum_{i=1}^{N}(\lambda_i\xi_i + \lambda'_i\xi'_i) - \sum_{i=1}^{N}(\eta_i\zeta_i + \eta'_i\zeta'_i) \\
&+ \sum_{i=1}^{N}\alpha_i\big[Re\left(y_i - \boldsymbol{w}^T\boldsymbol{x}_i - b\right) - \varepsilon - \xi_i\big] \\
&+ \sum_{i=1}^{N}\alpha'_i\big[Re\left(-y_i + \boldsymbol{w}^T\boldsymbol{x}_i - b\right) - \varepsilon - \xi'_i\big] \\
&+ \sum_{i=1}^{N}\beta_i\big[Im\left(y_i - \boldsymbol{w}^T\boldsymbol{x}_i - b\right) - j\varepsilon - j\zeta_i\big] \\
&+ \sum_{i=1}^{N}\beta'_i\big[Im\left(-y_i + \boldsymbol{w}^T\boldsymbol{x}_i + b\right) - j\varepsilon - j\zeta'_i\big]
\end{aligned}
\tag{4.3}
$$

Besides, the KKT conditions force $\frac{\partial L_{pd}}{\partial \boldsymbol{w}} = 0$, $\frac{d L_{pd}}{d b} = 0$, $\lambda_n\xi_n = 0$, $\lambda'_n\xi'_n = 0$ and $\eta_n\zeta_n = 0$, $\eta'_n\zeta'_n = 0$. Applying them, we obtain an optimal solution for the SVM regressor weights

$$
\boldsymbol{w} = \sum_{i=1}^{N}\psi_i^*\boldsymbol{x}_i^*
\tag{4.4}
$$

where $\psi_n = \alpha_n - \alpha'_n + j(\beta_n - \beta'_n)$. This result is analogous to the one for the real-valued SVM regressor problem, except that now Lagrange multipliers α_i and β_i for both the real and the imaginary components have been considered.

The norm of the complex-valued coefficients can be written as

$$||\boldsymbol{w}||^2 = \sum_{i=1}^{N} \sum_{j=n_0}^{N} \psi_i^* \psi_j \boldsymbol{x}_i^H \boldsymbol{x}_j = \psi^H \boldsymbol{R} \psi \qquad (4.5)$$

where $\boldsymbol{R}_{ij} = \boldsymbol{x}_i^H \boldsymbol{x}_j$.

By substituting (4.4) in functional (4.3) and proceeding as in the real-valued case to reduce the functional, one arrives to the dual functional

$$L_d = -\frac{1}{2}\psi^H \left(\boldsymbol{R} + \frac{\gamma}{2}\boldsymbol{I}\right) \psi + Re[\psi^H \boldsymbol{y}] - (\alpha + \alpha' + \beta + \beta')\mathbf{1}\varepsilon \qquad (4.6)$$

where we introduced a numerical regularization term $\frac{\gamma}{2}\boldsymbol{I}$ as we did in the real-valued case.

Expression (4.6) is formally equal to the real-valued expression SVM regression functional *(2.21)*. This is still a quadratic form that can be solved using the same strategies suitable for the real-valued Support Vector Machines.

The functional for SVM-autoregressive (ARx) models has exactly the same form [35]. For SVM classifiers, the expression for the functional is simply

$$L_d = -\frac{1}{2}\psi^H \boldsymbol{Y}^H[\boldsymbol{R} + \gamma \boldsymbol{I}]\boldsymbol{Y}\psi + \psi \qquad (4.7)$$

where for this problem $\psi_i = \alpha + j\beta$

4.2   LINEAR SUPPORT VECTOR ARx

Let $\{y_n\}$ be a discrete time series, from which a set of N consecutive samples are measured and grouped in a vector of observations,

$$\mathbf{y} = [y_1, y_2, \ldots, y_N]^T \qquad (4.8)$$

and let the set of vectors $\{\mathbf{z}^p\}$ be a base expanding a P-dimensional subspace into an N-dimensional space, given by the following P vectors

$$\mathbf{z}^p = \left[z_1^p, z_2^p, \ldots, z_N^p\right]^T, \quad p = 1, \ldots, P \qquad (4.9)$$

Each observed signal vector $\mathbf{y}$ can be represented as a linear combination of elements of this base, plus an error term $\mathbf{e} = [e_1, \ldots, e_N]^T$ modeling the measurement errors, as follows:

$$\mathbf{y} = \sum_{p=1}^{P} w^p \mathbf{z}^p + \mathbf{e} \qquad (4.10)$$

For a given time instant n, a linear time series model can be written as

$$y_n = \sum_{p=1}^{P} w^p z_n^p + e_n = \boldsymbol{w}^T \mathbf{v}_n + e_n \qquad (4.11)$$

where $\boldsymbol{w} = [w^1, \ldots, w^P]^T$ is the model weight vector to be estimated, and $\mathbf{v}_n = [z_n^1, \ldots, z_n^P]^T$ represents the *input space* at time instant n.

Note that the input space is closely related to the signal space. The input vector is given at time instant n by the nth element of the signal space vectors. For instance, in the case of a nonparametric spectral estimation, the base is formed by the lagged sinusoidal harmonics [36]. In the case of ARx system identification and modeling, the base in space is generated by the input signal and the delayed versions of input and output signals [35].

The expression in Eq. *(4.11)* is the one of a regressor and the model parameters can be adjusted using the SVR optimization described in Chapter 2. Applying it lead to the following expression for the filter coefficients

$$w^p = \sum_n (\alpha_n - \alpha_n^*) z_n^p \qquad (4.12)$$

and to the dual functional of Eq. *(2.20)*,

$$-\frac{1}{2}(\boldsymbol{\alpha} - \boldsymbol{\alpha}^*)^T [\boldsymbol{R}_v + \delta \mathbf{I}] (\boldsymbol{\alpha} - \boldsymbol{\alpha}^*) + (\boldsymbol{\alpha} - \boldsymbol{\alpha}^*)^T \mathbf{y} - \varepsilon \mathbf{1}^T (\boldsymbol{\alpha} + \boldsymbol{\alpha}^*) \qquad (4.13)$$

constrained to

$$C \geq \alpha_n^{(*)} \geq 0, \qquad (4.14)$$

where $\boldsymbol{\alpha}^{(*)} = \left[\alpha_1^{(*)}, \cdots, \alpha_N^{(*)} \right]^T$, in whose formulation we can identify the cross-correlation matrix of input space vectors and denote it as

$$\boldsymbol{R}_v(s, t) \equiv \mathbf{v}_s^T \mathbf{v}_t \qquad (4.15)$$

After obtaining Lagrange multipliers $\boldsymbol{\alpha}^{(*)}$, the time series model for a new sample at time instant m can be readily expressed as

$$y_m = f(\mathbf{v}_m) = \sum_{n=1}^{N} (\alpha_n - \alpha_n^*) \mathbf{v}_n^T \mathbf{v}_m, \qquad (4.16)$$

which is a function of weights in the input space associated to nonzero Lagrange multipliers. Hence, identification of a suitable base in the space becomes crucial but also straightforward. A general framework for nonlinear versions of SVM-ARMA models can be found in [37].

4.2.1 Support Vector Auto-Regressive Model

A simple way to estimate the spectrum of a signal is to compute a model of the signal then compute the spectrum of the model (cf. [38]). The observed signal is modeled as the output of a linear system driven by white noise, and the coefficients of the linear system are estimated by Yule–Walker, Burg, or Covariance methods [39]. The most used linear system model is the all-pole structure, and output y_n of such a filter for a white noise input is an auto-regressive (AR) process of order P, AR(P), which can be expressed by a difference equation

$$y_n = \sum_{p=1}^{P} w^p y_{n-p} + e_n \tag{4.17}$$

where w^p are the AR parameters, and e_n denotes the samples of the innovation process. Once the coefficients w^p of the AR process are calculated, the PSD estimation can be computed. Examples of application are shown in Subsection 6.1.1.

4.2.2 SVM-ARMA Complex Formulation

Let us consider now two complex, discrete time processes $\{x_n\}$ and $\{y_n\}$, representing the input and the output, respectively, of a linear, time invariant system that is defined by the following finite difference equation

$$y_n = \sum_{i=1}^{M} a^i y_{n-i} + \sum_{j=1}^{Q} b^j x_{n-j+1} + e_n \tag{4.18}$$

where $\{a^i\}$, $\{b^j\}$ are M and Q complex coefficients standing for the autoregressive (AR) and the moving-average (MA) parts of the system, respectively, and e_n is complex residual noise. Model (4.18) can be rewritten into its vector form, as given in (4.11), by introducing the following notation:

$$\mathbf{z}^p = [y_{1-p}, \ldots, y_{N-p}, x_p, \ldots, x_{N-p+1}]^T$$
$$\mathbf{w} = [\mathbf{a}^T, \mathbf{b}^T]^T \tag{4.19}$$

Therefore, the space base vectors are formed by delayed versions of both the output and the input time series, and again, initial conditions are necessary.

Once the base in the space has been identified, we must check the validity of the proposed cost function and the functional to be minimized. In the two previous examples, these steps are trivially obtained. Nevertheless, for complex SVM-ARMA modeling, we must include real and imaginary slack variables in the cost function and their corresponding constraints as in (4.2). Introducing the constraints into the primal functional by means of Lagrange multipliers $\{\alpha_{R,n}\}$, $\{\alpha^*_{R,n}\}$, $\{\alpha_{I,n}\}$, $\{\alpha^*_{I,n}\}$, for the real (subscript R) and imaginary (subscript I) parts, we obtain the primal-dual function. By making zero its gradient with respect to w^p,

$$\mathbf{a} = \sum_{n=n_0}^{N} \psi_n \mathbf{y}^*[n-1]$$

$$\mathbf{b} = \sum_{n=n_0}^{N} \psi_n \mathbf{x}^*[n]$$

(4.20)

where $\psi_n = (\alpha_{R,n} - \alpha_{R,n}^*) + j(\alpha_{I,n} - \alpha_{I,n}^*)$. The analytical relationship between the residuals and the Lagrange multipliers is identical to that of the real case.

Placing the optimal solution (4.20) in the primal-dual functional and grouping terms lead to a compact form of the functional problem, which is equal to the one of Eq. *(4.7)*

$$-\frac{1}{2} \psi^H (\mathbf{R}_{\mathbf{z}} + \delta \mathbf{I}) \psi + Re\left(\psi^H \mathbf{y}\right) - (\alpha_R + \alpha_R^* + \alpha_I + \alpha_I^*)\mathbf{1}\varepsilon$$

which where the Gram matrix $\mathbf{R}$ of dot products has the expression

$$\mathbf{R}_{\mathbf{z}}(s, t) \equiv \mathbf{z}_s^T \mathbf{z}_t$$

(4.21)

4.3   ROBUST COST FUNCTION OF SUPPORT VECTOR REGRESSORS

In Eqs. *(2.21)* and *(4.7)* we introduced a term $\frac{\gamma}{2}\mathbf{I}$ to avoid numerical problems in those situations in which the eigenvectors of the matrix are too small. In such a situation, the optimization problem is ill-posed, so adding a small diagonal matrix increases the lowest eigenvectors without significantly biasing the solution.

We present here another interpretation of the effect of this small diagonal term as a modification of the ε-insensitive cost function *(2.14)* [10]. This interpretation will allow us to use other criteria to adjust the value of γ depending of the nature and power of the noise present in the data.

To do that, we simply modify the ε-insensitive cost function and repeat the optimization computations of the resulting primal functional. The modified cost function is

$$L(\xi) = \begin{cases} 0 & |\xi| \leq \varepsilon \\ \frac{1}{2\gamma}(|\xi| - \varepsilon)^2 & \varepsilon \leq |\xi| \leq e_C \\ C(|\xi| - \varepsilon) - \frac{1}{2}\gamma C^2 & |\xi| \geq e_C \end{cases}$$

(4.22)

where $e_C = \varepsilon + \gamma C$; ε is the insensitive parameter, and γ and C control the trade-off between the regularization and the losses. Three different regions allow to deal with different kinds of noise (see Fig. 4.2): ε-insensitive zone ignores errors lower than ε; quadratic cost zone uses the L_2-norm of errors, which is appropriate for Gaussian noise; and linear cost zone limits

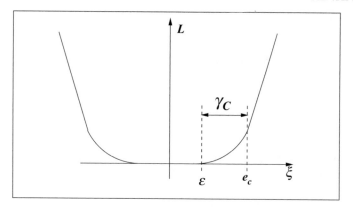

FIGURE 4.2: Cost function with ε-insensitive, quadratic, and linear cost zones ($e_C = \varepsilon + \gamma C$)

the effect of outliers. Note that *(4.22)* represents Vapnik ε-insensitive cost function when γ is small enough, and Huber's cost function [40] when $\varepsilon = 0$. γ also plays the role of numerical regularization in the quadratic problem.

In most situations, the noise present at the signal output is the sum of thermal noise plus other contributions coming from external interference, which are usually sub-Gaussian. In these situations, the probability density function (PDF) is the convolution of both PDFs, but a good approximation near the origin is the Gaussian one. Then, a L_2 cost function will give the maximum-likelihood (ML) estimator. Far from the origin, the sub-Gaussian approach is a more convenient one. Then, a L_ρ, $\rho < 1$ (sub-linear) cost function will yield the ML estimator. Nevertheless, to assure convergence, the cost function must be convex. The lowest degree in the cost function which leads to a convex optimization problem is 1, so the most convenient cost function far from the origin is linear.

The primal which fits the loss function *(4.22)* is

$$L_p = \frac{1}{2}\|\boldsymbol{w}\|^2 + \frac{1}{2\gamma}\sum_{i\in I_1}\left(\xi_i^2 + \xi_i'^2\right) + C\sum_{i\in I_2}\left(\xi_i + \xi_i' - \frac{\gamma C}{2}\right) \qquad (4.23)$$

I_1, I_2 are the sets of samples for which losses are required to have a quadratic or a linear cost, respectively. These sets are not necessarily static during the optimization procedure. The constraints are the same as in *(2.13)*

$$-y_i + \boldsymbol{w}^T\boldsymbol{x}_i + b + \xi_i + \varepsilon \geq 0$$
$$y_i - \boldsymbol{w}^T\boldsymbol{x}_i - b + \xi_i' + \varepsilon \geq 0$$
$$\xi_i, \xi_i' \geq 0$$

Again, we must apply the technique of Lagrange multipliers to solve the constrained optimization problem. The details of the derivation are provided in Appendix C.

The resulting expression for the dual functional is

$$L_d = -\frac{1}{2}\sum_{i=1}^{N}\sum_{i=1}^{N}(\alpha_i - \alpha'_i)\boldsymbol{x}_i^T\boldsymbol{x}_j(\alpha_j - \alpha'_j)$$
$$+ \sum_{i=1}^{N}\left((\alpha_i - \alpha'_i)y_i - (\alpha_i + \alpha'_i)\varepsilon - \frac{\gamma}{2}(\alpha_i^2 + \alpha'^2_i)\right)$$

(4.24)

A matrix expression of this equation can be easily found by rearranging terms and taking into account that $\alpha_i\alpha'_i = 0$

$$L_d = -\frac{1}{2}(\boldsymbol{\alpha} - \boldsymbol{\alpha}')^T[\boldsymbol{R} + \gamma\boldsymbol{I}](\boldsymbol{\alpha} - \boldsymbol{\alpha}') + (\boldsymbol{\alpha} - \boldsymbol{\alpha}')^T\boldsymbol{y} - (\boldsymbol{\alpha} + \boldsymbol{\alpha}')\boldsymbol{1}\varepsilon$$

In this equation the numerical regularization term is γ. This parameter has an interpretation in terms of the applied cost function. Applying it is equivalent to apply a cost function which has quadratic interval between ε and $\varepsilon + \gamma C$. If the motivation for the use of γ is to raise the values of the eigenvectors of $\boldsymbol{R}$, then, its value is very low small, and the quadratic interval is negligible. Nevertheless, greater values of γ can also be used in order to make the interval wide enough to include all the data which is mainly corrupted by additive white Gaussian noise (AWGN), letting the linear interval to the outliers, or samples corrupted by impulse or other sub-Gaussian noises.

4.4   PARAMETER SELECTION

The trade-off between generalization ability and empirical risk minimization in an SVM is controlled by the parameter C, as seen in Chapter 2 (see also Appendix A). This parameter may eventually be critical, for example, if the sample set for training is too small. The user may choose to sweep this parameter using a validation set to find the best value for this parameter, but this is sometimes impossible due to a lack of data or due to the fact that this procedure is time consuming and some applications cannot afford the delay that such a search of C implies. The same occurs with the parameter ε, which allows a sparse solution for SVM regressors, but may produce poor results if it is not carefully chosen.

The used kernels also have parameters, whose choice can make the difference between an SVM with reasonable results and an SVM with poor ones.

There are several criteria to choose these parameters. In order to choose C and the kernel parameters for classification, we summarize the Kwok's criterion based on the Evidence Framework [41]. To choose C and ε in regression, we summarize the Evidence Framework criterion by Kwok [42] and the criterion of Cherkassky [43].

4.4.1 The Evidence Framework Criterion

The Evidence Framework [44] is applied in [41] to find the parameters of a nonlinear SVM for classification. In this framework a classifier is assumed to have a classification function $y = f(\boldsymbol{w}, \boldsymbol{x})$, a distribution $p(\boldsymbol{x}|\boldsymbol{w}, f)$ and a prior parameter distribution

$$p(\boldsymbol{w}|f, \lambda) = \frac{\exp(-\lambda E_W(\boldsymbol{w}, f))}{Z_W(\lambda)} \qquad (4.25)$$

where λ is some regularization parameter, and Z_W is a normalization parameter for (4.25) to have distribution properties. The method first obtains a posterior distribution of the parameters (level 1 inference) using (4.25) and applying the Bayes rule:

$$p(\boldsymbol{w}|\boldsymbol{x}, f, \lambda) = \frac{\exp(-M(\boldsymbol{w}))}{Z_W(\lambda)} \qquad (4.26)$$

where $M(\boldsymbol{w}) = \lambda E_W(\boldsymbol{w}) - \log p(\boldsymbol{x}|\boldsymbol{w}, f)$. The maximization of (4.26) (minimization of $M(\boldsymbol{w})$) leads to the maximum a posteriori (MAP) value $\boldsymbol{w}_{MAP}$.

The distribution (4.26) can be approximated by a Gaussian whose mean value is $M(\boldsymbol{w}_{MAP})$ and whose covariance matrix is the Hessian $\boldsymbol{A} = \nabla^2 M$. It can be seen [44] that

$$\log p(\boldsymbol{x}|\lambda, f) = -\lambda E_W^{MAP} + G^{MAP} - \frac{1}{2} \log \det(\boldsymbol{A}) + \frac{k}{2} \log(\lambda) \qquad (4.27)$$

where $E_W^{MAP} = E_W(\boldsymbol{w}_{MAP}|f)$ and $G^{MAP} = \log p(\boldsymbol{x}|\boldsymbol{w}_{MAP}, f)$. The quantity (4.27) is the so-called evidence of λ. In order to iteratively obtain the MAP value for λ (level 2 inference), the derivative of its evidence (4.27) has to be set to zero. Then, the following result is obtained

$$2\lambda_{MAP} E_W^{MAP} = \gamma \qquad (4.28)$$

where $\gamma = k - \lambda \text{trace}(\boldsymbol{A}^{-1})$, k being the dimension of $\boldsymbol{w}$. Then, going to (4.26), $\boldsymbol{w}_{MAP}$ has to be recalculated and the process iterated until convergence.

Finally, the model f can be ranked by examining its posterior probability $p(f|\boldsymbol{x}) = p(\boldsymbol{x}|f)p(\boldsymbol{x})$ (level 3 inference), taking into account that using a Gaussian approximation, $p(\boldsymbol{x}|f) = p(\boldsymbol{x}|\lambda_{MAP}, f)/\sqrt{\gamma}$.

4.4.2 Selection of ε for Gaussian and Exponential Noises

In [42], the distribution function is introduced for the output y_i of an SVM regression model in which the ε-insensitive cost function has been applied. This distribution has the form

$$p(y_i|\boldsymbol{x}_i, \boldsymbol{w}, \beta, \varepsilon) = \frac{\beta}{2(1 + \varepsilon\beta)} \exp(-\beta|y_i - \boldsymbol{w}^T \boldsymbol{x}_i|) \qquad (4.29)$$

Using a method similar to the presented in the previous section, an expression for the MAP value of $\boldsymbol{w}$ is obtained, then an evidence for ε can be computed. Assuming that the noise affecting y_i

is Gaussian, a closed expression for the optimum value of ε is computed

$$\varepsilon = 1.0043\sigma \qquad (4.30)$$

where σ^2 is the noise power. Also, a computation of ε is done for exponential noise, whose result is $\varepsilon = 0$.

4.4.3   Selection of C and ε Based on Heuristics

The previous parameter selection criterion is based in a model of prior probability of the output y_i which does not take into account the sample size, but that may be inaccurate for those cases in which the sample size is small, as the models are asymptotic with the number of samples. Also, the possible presence of outliers is not taking into account into these criterion. Cherkassky [43] introduced an alternative method for selecting ε and C in regression which takes into account the sample size and it is robust against outliers. Taking into account the SVM expression for f

$$f(\boldsymbol{x}_i) = \sum_j \alpha_j K(\boldsymbol{x}_i, \boldsymbol{x}_j)$$

and knowing that $\alpha_j \leq C$, the following inequality can be derived for kernels which satisfy $K(\cdot, \cdot) \leq 1$

$$|f(\boldsymbol{x}_i)| \leq n_{sv}C \qquad (4.31)$$

where n_{sv} is the number of support vectors. This number inversely depends on the value of ε. The expression is meaningful, as it suggests us that the variance of the output may increase if we increase C or if we decrease ε. Cherkassky suggests to use the heuristic value

$$C = 3\sigma \qquad (4.32)$$

It is known that the noise standard deviation of the output of a linear regressor is proportional to the input noise and inversely proportional to the number of training samples. Also, it is known [42] that the value of ε should be proportional to the input noise. Then, in [43] an heuristic is applied for the selection of it:

$$\varepsilon = 3\sigma\sqrt{\frac{\ln N}{N}} \qquad (4.33)$$

This heuristic provides better performance than (4.30) when the number of samples is small. When the number of samples is high, this heuristic value tends to zero, so (4.30) performs better.

CHAPTER 5

Support Vector Machines for Beamforming

Support Vector Machines (SVM) have shown clear advantage in prediction, regression and estimation over classical approaches in a wide range of applications due to its improved generalization performance, which is important when small datasets are available for the training of the algorithms. Array signal processing involves complex signals, for which a complex-valued formulation of the SVM is needed. We introduced this formulation by introducing the real and imaginary parts of the error in the primal optimization and then proceeding as usual to solve a complex-valued constrained optimization problem in the previous chapter. The resulting algorithm is a natural counterpart of the real-valued Support Vector Regressor, which can be immediately applied to array signal processing.

In this chapter, we will make use of the SVM regression framework presented in the previous chapter to array beamforming. The first section is devoted to the application of the SVM to linear beamforming. The second one deals with the application of nonlinear Support Vector Regressors to the fast estimation of linear beamformer parameters, and the last section develops strategies to construct nonlinear beamformers.

5.1 PROBLEM STATEMENT

The signal received by an array consists of the desired or friendly signal, which may come from one or more directions of arrival with different amplitudes and phases and the interfering or undesired signals (Fig. 5.1).

The output vector of an M-element array receiving K signals can be written in matrix notation as

$$x[\mathbf{n}] = \mathbf{A}s[n] + \mathbf{g}[n] \tag{5.1}$$

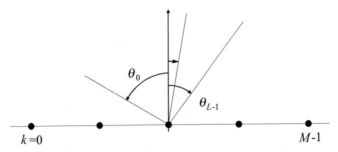

FIGURE 5.1: Angles of arrival

where

$$\mathbf{A} = [\mathbf{a}(\theta_1) \cdots \mathbf{a}(\theta_k)]$$
$$\mathbf{a}(\theta_i) = \left[1e^{-jk_i} \cdots e^{-j(M-1)k_i}\right]^T$$
$$\mathbf{s}[n] = [s_1[n] \cdots s_k[n]]^T$$
$$\mathbf{g}[n] = [g_1[n] \cdots g_M[n]]^T$$

$$(5.2)$$

are respectively the $M \times K$ steering matrix and vectors, the received signals and the thermal noise present at the output of each array element.

The spatial correlation matrix of the received signals is

$$\mathbf{R} = \frac{1}{N}\mathbf{X}\mathbf{X}^H$$
$$= E[(\mathbf{A}\mathbf{s}[n] + \mathbf{g}[n])(\mathbf{A}\mathbf{s}[n] + \mathbf{g}[n])^H]$$
$$= \mathbf{A}\mathbf{P}\mathbf{A}^H + \sigma_g^2 I$$

$$(5.3)$$

provided that the signal and noise are independent, where $\mathbf{P}$ is the autocorrelation of $\mathbf{s}[n]$, and σ_g^2 is the noise power.

5.1.1   Temporal Reference

Assuming that certain transmitted data is known for training purposes, one may apply the Minimum Mean Square Error (MMSE) criterion to the received signal, that is

$$\min_{\boldsymbol{w}} E\{\|e\|^2\}$$

$$(5.4)$$

where

$$e[n] = \hat{y}[n] - \boldsymbol{w}^H\mathbf{x}[n]$$

$$(5.5)$$

which leads to the well-known and widely used Wiener solution

$$\boldsymbol{w} = \mathbf{R}^{-1}\mathbf{p}$$

$$(5.6)$$

$\mathbf{p}$ being the cross-correlation between the desired output and the incoming signal.

5.1.2   Spatial Reference

If the AOA of the desired signal is known, the optimal beamformer needs to satisfy

$$\min\{\boldsymbol{w}\mathbf{R}\boldsymbol{w}^H\} \tag{5.7}$$

subject to the constraint

$$\boldsymbol{w}^H\mathbf{s}_d[n] = y[n] \tag{5.8}$$

That is, it minimizes the output power subject to the condition that the desired signal $\mathbf{s}_d[n]$ produces a given output $y[n]$. This is the classical solution to beamforming with spatial reference, also known as Minimum Variance Distortionless Method (MVDM), as it assumes the use of the covariance method of data windowing [45]. Applying Lagrange multipliers to that constrained optimization problem leads to the solution

$$\boldsymbol{w}_{opt} = \mathbf{R}^{-1}\mathbf{S}_d[\mathbf{S}_d^H\mathbf{R}^{-1}\mathbf{S}_d]^{-1}\mathbf{y} \tag{5.9}$$

where $\mathbf{S}_d$ is the matrix of all desired input vectors, and $\mathbf{y}$ is the vector of all desired output signals. It is straightforward to prove that both MMSE and MVDM lead to the same solution, provided that the autocorrelation matrix $\mathbf{R}$ is accurately computed.

5.2   LINEAR SVM BEAMFORMER WITH TEMPORAL REFERENCE

The output vector $\mathbf{x}[\mathbf{n}]$ is linearly processed to obtain the desired output $d[n]$. The expression for the output of the array processor is

$$y[n] = \boldsymbol{w}^T\boldsymbol{x}[n] = d[n] + \epsilon[n] \tag{5.10}$$

where $\mathbf{w} = [w_1 \cdots w_M]$ is the weight vector of the array and $\epsilon[n]$ is the estimation error.

For a set of N observed samples of $\{x[n]\}$ and when nonzero empirical errors are expected, the functional to be minimized is

$$\frac{1}{2}||\boldsymbol{w}||^2 + \sum_{n=1}^{N} L_R(e[n], \varepsilon, \gamma, C) \tag{5.11}$$

Therefore, particularizing to the error cost function (4.22), we have to minimize [46]

$$\frac{1}{2}||\boldsymbol{w}||^2 + \sum L_R(\xi_n + \xi_n') + \sum L_R(\zeta_n + \zeta_n') \tag{5.12}$$

subject to

$$
\begin{aligned}
Re\left(d[n] - \boldsymbol{w}^T\boldsymbol{x}[n]\right) &\leq \varepsilon + \xi_n \\
Re\left(-d[n] + \boldsymbol{w}^T\boldsymbol{x}[n]\right) &\leq \varepsilon + \xi'_n \\
Im\left(d[n] - \boldsymbol{w}^T\boldsymbol{x}[n]\right) &\leq \varepsilon + \zeta_n \\
Im\left(-d[n] + \boldsymbol{w}^T\boldsymbol{x}[n]\right) &\leq \varepsilon + \zeta'_n \\
\xi[n], \xi'[n], \zeta[n], \zeta'[n] &\geq 0
\end{aligned}
\tag{5.13}
$$

where $\xi[n]$ ($\xi'[n]$) stand for positive (negative) errors in the real part of the output -analogously for $\zeta[n]$ ($\zeta'[n]$) in the imaginary part. As in other SVM formulations, parameter C can be seen as a trade-off factor between the empirical risk and the structural risk.

Now the functional in Eq. *(4.6)* can be applied

$$
L_d = -\frac{1}{2}\psi^H Re\left(\boldsymbol{R} + \frac{\gamma}{2}\boldsymbol{I}\right)\psi + Re[\psi^H\mathbf{y}] - (\alpha + \alpha' + \beta + \beta')\mathbf{1}\varepsilon
\tag{5.14}
$$

The optimization of this functional will give a set of parameters $\psi_i = \alpha_i + j\beta_i$, and using expression *(4.4)*

$$
\boldsymbol{w} = \sum_{i=1}^{N} \psi_i^* \boldsymbol{x}_i^*
$$

we can compute the weights of the beamformer. Combining this equation and *(5.10)*, the beamformer can be expressed as

$$
y[n] = \sum_{i=1}^{N} \psi_i^* \boldsymbol{x}_i^{*T} \boldsymbol{x}[n]
\tag{5.15}
$$

5.2.1   Bit Error Rate Performance

We first test the algorithm against to the regularized Least Squares (LS) method [45] with an array of 6 elements. The desired signal comes from the angles of arrival (AOA) -0.1π and 0.25π, with amplitudes 1 and 0.3, and the interfering signals come from the AOAs -0.05π, 0.1π and 0.3π with amplitude 1.

In order to train the beamformer, a burst of 50 known symbols is sent. Then, the Bit Error Rate (BER) is measured with bursts of 10,000 unknown symbols.

In this and in the next examples, we fix γ of Eq. *(4.6)* to 10^{-6} and then the product γC is adjusted to the noise standard deviation. That way, most of the samples which are corrupted by thermal noise will fall in the quadratic area, where the outliers produced by interfering signals will fall in the linear area.

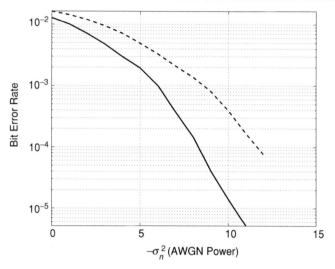

FIGURE 5.2: BER performance for experiment 1. SVM (continuous line) and regularized LS (dashed line) beamformers. (Source [46]. Reprinted with permission of the IEEE)

We calculate the BER performance of LS method and SVM for different noise levels from 0 dB to −15 dB. Each BER has been measured by averaging the results of 100 independent trials. The results can be seen in Fig. 5.2.

We repeat the same example with the desired signals coming from the angles of arrival (AOA) -0.1π and 0.25π, with amplitudes 1 and 0.3, and the interfering signals coming from the AOAs $-0.02\pi, 0.2\pi, 0.3\pi$ with amplitude 1 (see Fig. 5.3). In this example, the interfering

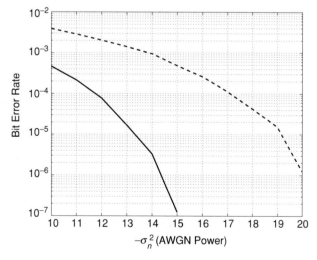

FIGURE 5.3: BER performance for experiment 2. SVM (continuous line) and regularized LS method (dashed line) beamformers. (Source [46]. Reprinted with permission of the IEEE)

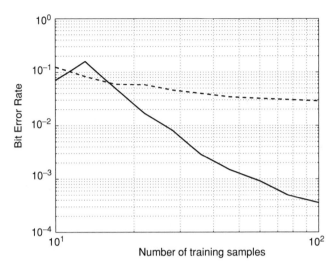

FIGURE 5.4: BER performance against the number of training samples. SVM (continuous line) and regularized LS method (dashed line) beamformers. (Source [46]. Reprinted with permission of the IEEE)

signals are much closer to the desired ones, thus biasing the LS method algorithm. The superior performance of SVM is due to its better robustness against the non-Gaussian outliers produced by the interfering signals.

5.2.2   Robustness Against Overfitting

One advantage of the SVM is that its generalization ability is controlled by the regularization imposed by the minimization of the weight vector norm. We highlight this fact by measuring the Bit Error Rate for different number of training samples. The results are shown in Fig. 5.4.

5.3   LINEAR SVM BEAMFORMER WITH SPATIAL REFERENCE

In [47, 48], an algorithm has been introduced that uses the knowledge of the AOA of the desired signal in a similar way as in *(5.7)*. The procedure consists of a transformation of the solution provided by the SVM. In [49, 50], a variant has been introduced that uses an explicit form of the complex-valued SVM and that includes a nonlinear version of the algorithm (see 5.6).

A linear SVM beamformer with sidelobe control can be expressed as

$$L_p = \frac{1}{2}\boldsymbol{w}^H \mathbf{R}\boldsymbol{w} + \sum_i L(\xi_i + \xi_i') + \sum_i L(\zeta_i + \zeta_i') \qquad (5.16)$$

where L is the cost function *(4.22)* applied to the MVDM constraints. These constraints have been adapted to the Support Vector formulation as follows:

$$\mathbb{R}e\left(y[n] - \boldsymbol{w}^H \mathbf{s}_d[n]\right) \leq \varepsilon - \xi[n]$$
$$\mathbb{I}m\left(y[n] - \boldsymbol{w}^H \mathbf{s}_d[n]\right) \leq \varepsilon - \zeta[n]$$
$$\mathbb{R}e\left(-y[n] + \boldsymbol{w}^H \mathbf{s}_d[n]\right) \leq \varepsilon - \xi'[n]$$
$$\mathbb{I}m\left(-y[n] + \boldsymbol{w}^H \mathbf{s}_d[n]\right) \leq \varepsilon - \zeta'[n]$$

$$(5.17)$$

Applying Lagrange analysis to this primal functional straightforwardly gives the result

$$\boldsymbol{w} = \mathbf{R}^{-1} \mathbf{S}_d \boldsymbol{\psi} \tag{5.18}$$

Applied to the primal, the previous result leads to the dual

$$-\frac{1}{2}\boldsymbol{\psi}^H \left[\mathbf{S}_d^H \mathbf{R}^{-1} \mathbf{S}_d + \gamma \mathbf{I}\right] \boldsymbol{\psi} - \mathbb{R}e(\boldsymbol{\psi}^H \mathbf{y}) + \varepsilon \mathbf{1}(\boldsymbol{\alpha} + \boldsymbol{\beta} + \boldsymbol{\alpha}' + \boldsymbol{\beta}') \tag{5.19}$$

where $\mathbf{y} = [y[1], \cdots, y[N]]$. Note that a regularization term naturally appears from the application of the cost function (see Section 4.3).

5.4   NONLINEAR PARAMETER ESTIMATION OF LINEAR BEAMFORMERS

The inconvenient of linear beamformers is that the parameter vector has to be recalculated each time the direction of arrival of the sources or the number of sources changes. The parameter vector $\mathbf{w}$ can be estimated using a nonlinear regressor so that it can place nulls in the direction of the interfering signals and track the desired signal. The difference between the linear and the nonlinear approaches is that the last one *can be trained off line with a large number of different angles of arrival of signals* [51].

Next, during detection, the network computes the beamformer by performing a nonlinear interpolation. That way, the network does not need to be retrained, as it contains the needed information to estimate the optimal beamformer. The adaptation procedure becomes then faster.

5.4.1   Generation of the Training Data

We can assume that a dataset consisting of L matrices of the form $\mathbf{X}^l = \{\mathbf{x}_1^l \cdots \mathbf{x}_N^l\}$, $1 < l < L$ is generated using the definition *(5.1)*. In order to make the network robust to changes in amplitude and to make it less dependent of its kernel parameters, the data has to be normalized in terms of amplitude. One effective way is simply to divide them by its norm.

The information that we need to compute the beamformer is present in the autocorrelation matrix of the data. As in the training phase the data is artificially generated and the autocorrelation matrices of all samples can be exactly computed by

$$\mathbf{R}^l = \mathbf{A}^l \mathbf{P}^l (\mathbf{A}^l)^H + \left(\sigma_g^l\right)^2 I$$

They have to be normalized to have diagonal elements equal to one, as this is the power of the assumed received data.

The next step is to produce the optimal parameter vector for the beamformer using expression *(5.9)*. Then, a set of pairs $\{\mathbf{X}^l, \mathbf{w}^l\}$ is available to train a set of M nonlinear regressors, M being the number of elements of $\mathbf{w}$.

5.4.2 Structure of the Estimator

There are two possible approaches for the training of the regressors. The structure is depicted in Fig. 5.5. The input consists of the signal vector $\{x_1[n] \cdots x_M[n]\}$, and the outputs are the elements of the parameter vector w.

Taking into account that input and output should be complex-valued, the output of regressor m, $1 \leq m \leq M$, is equivalent to that of *(4.4)*

$$w_m = \sum_{i=1}^{N} \psi_{i,m} K(\mathbf{x}_i, \mathbf{x}_j) + b_m \qquad (5.20)$$

where $\psi_n = \alpha_n - \alpha_n' + j(\beta_n - \beta_n')$.

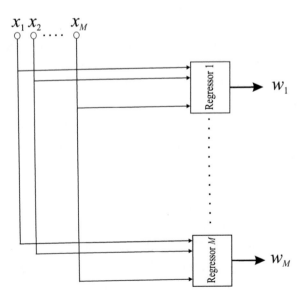

FIGURE 5.5: Structure of the multiregressor

The first approach is quite straightforward, and consists of independently training a number of regressors with each one of the M elements of $\boldsymbol{w}$. Then, each one will learn the particular behavior of each weight element. The performance of this approach is limited by the fact that each regressor does not use the information of the outputs of the others to train. But the outputs are not really independent, then this loss of information may decrease the performance. Implementation and training are exactly the ones presented in Subsection 3.1.5 and Section 4.1.

The other approach is to construct a multiregression structure in which we train all the outputs in a block. Then, the outputs are related, which may lead to a better performance. The expression of the output of the multiregressor is

$$\boldsymbol{w} = \varphi^H(\boldsymbol{x}_i)\boldsymbol{V} + \boldsymbol{b} + \boldsymbol{e}_i \qquad (5.21)$$

where

$$\begin{aligned} \boldsymbol{V} &= [\boldsymbol{v}_1 \cdots \boldsymbol{v}_M] \\ \boldsymbol{e}_i &= [e_{1,i} \cdots e_{N,i}] \end{aligned} \qquad (5.22)$$

The difference with the first approach is that here we compute all the parameters w_j in the same equation.

If we use this approach, we must generalize the ε-insensitivity cost function that we use in the optimization in *(2.14)*, which we reproduce here for convenience.

$$\ell(e_i) = \begin{cases} 0 & |e_i| < \varepsilon \\ |e_i| - \varepsilon & |e_i| > \varepsilon \end{cases}$$

That cost function is designed to allow errors less or equal to ε. As we explained in Subsection 2.2.2, the application of this function produces a sparse solution, in which the samples used to produce the optimal regressor are the subset of samples out of the ε band. This is valid when the output is a scalar. A first approach to generalize this cost function is to directly apply it to all the components of the regressor output, so the resulting primal functional is an extension of *(4.1)* to more dimensions, i.e.:

$$\begin{aligned} L_p &= \frac{1}{2}\sum_{m=1}^{M}||\boldsymbol{v}_m||^2 + C\sum_{m=1}^{M}\sum_{n\in I_1}(\xi_{n,m} + \xi'_{n,m}) + C\sum_{m=1}^{M}\sum_{n\in I_1}(\zeta_{n,m} + \zeta'_{n,m}) \\ &+ \frac{1}{2\gamma}\sum_{m=1}^{M}\sum_{n\in I_2}\left(\xi_{n,m}^2 + \xi'^2_{n,m}\right) + \frac{1}{2\gamma}\sum_{n\in I_2}\left(\zeta_{n,m}^2 + \zeta'^2_{n,m}\right) \end{aligned} \qquad (5.23)$$

subject to

$$
\begin{aligned}
Re\left(w_{n,m} - v_m^T \mathbf{x}_{n,m} - b_m\right) &\leq \varepsilon + \xi_{n,m} \\
Re\left(-w_{n,m} + v_m^T \mathbf{x}_{n,m} + b_m\right) &\leq \varepsilon + \xi'_{n,m} \\
Im\left(w_{n,m} - v_m^T \mathbf{x}_{n,m} - b_m\right) &\leq \varepsilon + \zeta_{n,m} \\
Im\left(-w_{n,m} + v_m^T \mathbf{x}_{n,m} + b_m\right) &\leq \varepsilon + \zeta'_{n,m} \\
\xi_{n,m}, \xi'_{n,m}, \zeta_{n,m}, \zeta'_{n,m} &\geq 0
\end{aligned}
\tag{5.24}
$$

This approach, however, is not the most suitable one due to the fact that the errors of the samples are not equally treated. The ε band turns out to be here a hypercube with M dimensions. The minimum distance between the origin and the hypercube is, obviously, ε, but the maximum distance is equal to the distance to one of the corners. This distance is $\sqrt{M}\varepsilon$. A sample whose error vector is close to the center of one of the faces of the hypercube will be discarded if the norm of the error vector exceeds ε. But for a sample whose error vector is close to a corner will not be discarded until its error norm exceeds $\sqrt{M}\varepsilon$ [52].

The solution provided in [52] equally treats all errors, as it use a quadratic approach. Let us denote the norm of the vector constructed with all errors as $||\mathbf{e}_i|| = \sqrt{\mathbf{e}_i^H \mathbf{e}_i}$. The cost function is expressed as

$$
\ell(e_i) = \begin{cases} 0 & ||\mathbf{e}_i|| < \varepsilon \\ (||\mathbf{e}_i|| - \varepsilon)^2 & ||\mathbf{e}_i|| > \varepsilon \end{cases}
$$

Then, instead of having a ε hypercube in the cost function, we will use a hypersphere. The primal functional is expressed as

$$
L_p = \frac{1}{2}\sum_{m=1}^M ||v_m||^2 + C\sum_{m=1}^M \ell(e_m)
\tag{5.25}
$$

This functional cannot be optimized using a quadratic programming approach. The alternative optimization introduced in [52] consists of an iterative weighted least squares (IWRLS) algorithm. This approach uses a quadratic approximation of the functional of the form

$$
L_p = \frac{1}{2}\sum_{m=1}^M ||v_m||^2 + \frac{1}{2}\sum_{m=1}^M a_m ||e_m||^2 + CT
\tag{5.26}
$$

where

$$
a_m = \begin{cases} 0 & |e|_{i,m} < \varepsilon \\ \dfrac{2C(|e|_{i,m}) - \varepsilon}{|e|_{i,m}} & |e|_{i,m} \geq \varepsilon \end{cases}
\tag{5.27}
$$

In each step of the recursion, it is necessary to compute the solution of functional *(5.26)*. It consists of computing its gradient and solving it for zero. The gradient computation leads to the set of equations

$$\begin{bmatrix} \varphi^H D_a \varphi + I & \varphi^H a \\ a^H \varphi & 1^T a \end{bmatrix} \begin{bmatrix} w_i \\ b_i \end{bmatrix} = \begin{bmatrix} \varphi^H D_a y_i \\ a^H y_i \end{bmatrix} \tag{5.28}$$

$1 \leq i \leq M$.

Let initialize parameters V_0 and b_0 to zero. Define V_k and b_k as the values of the parameters at iteration k. The procedure for the optimization has the following steps:

1. Obtain the solution $\hat{W}, \hat{b}$ to the functional *(5.26)*.
2. Compute

$$P_k = \begin{bmatrix} \hat{W} - W_k \\ \hat{b} - b_k \end{bmatrix} \tag{5.29}$$

3. Compute

$$\begin{bmatrix} W_{k+1} \\ b_k \end{bmatrix} = \begin{bmatrix} W_{k+1} \\ b_k \end{bmatrix} + \mu P_k \tag{5.30}$$

where μ is an adaptation step.

4. Compute e_i and a_i for the functional *(5.26)* and go back to step 1 until convergence.

5.4.3   Examples

A nonlinear SVM estimator with Gaussian kernel has been trained with different angles of arrival using the multiregressor described in Eqs. *(5.26)* to *(5.30)*. We use 50 different angles of arrival between $-80°$ and $80°$ and SNRs of 10 dB, 15 dB and 20 dB. For each SNR and angle of arrival we generate 10 different samples, thus the number of data for training is of 1500.

Figures 5.6 and 5.7 show the result of a test of the estimator using a set of ten snapshots with a DOA of 0 and 10 degrees respectively and an SNR of 10 db. Figures 5.8 and 5.9 are the result of the same test, but with an SNR of 20 dB.

In all figures we compare the result with those of the optimum MVDM beamformer. It can be observed that the difference is low, and it is comparable to the result of the application of the MV by estimating the covariance matrix R from a set of samples. Also, there is no appreciable difference between the results of the data with both SNRs.

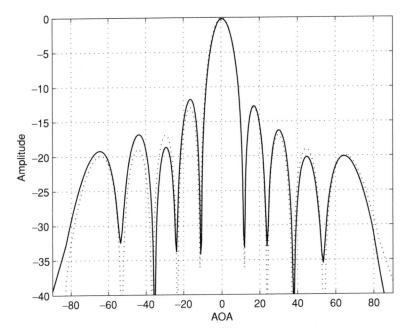

FIGURE 5.6: Beam diagrams for the SVM estimation of the parameters of a linear beamformer of 10 elements, with DOA at 0° and SNR = 10 dB (continuous line) and optimum MVDM (dotted line)

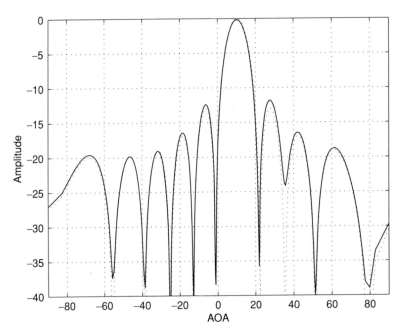

FIGURE 5.7: Beam diagrams for the SVM estimation of the parameters of a linear beamformer of 10 elements, with DOA at 10° and SNR = 10 dB (continuous line) and optimum MVDM (dotted line)

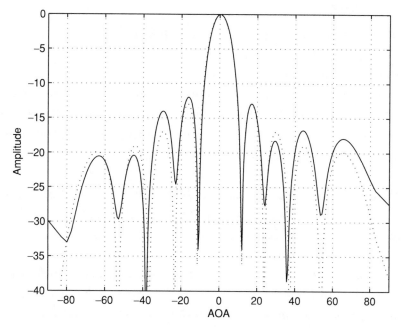

FIGURE 5.8: Beam diagrams for the SVM estimation of the parameters of a linear beamformer of 10 elements, with DOA at 0° and SNR = 20 dB (continuous line) and optimum MVDM (dotted line)

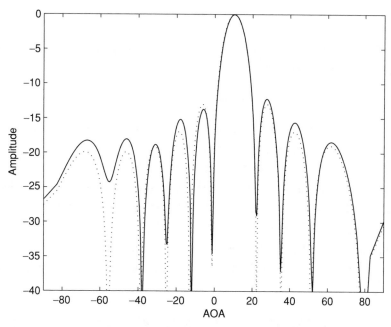

FIGURE 5.9: Beam diagrams for the SVM estimation of the parameters of a linear beamformer of 10 elements, with DOA at 10° and SNR = 20 dB (continuous line) and optimum MVDM (dotted line)

5.5   NONLINEAR SVM BEAMFORMER WITH TEMPORAL REFERENCE

5.5.1   Structure of the Estimator

The structure of a nonlinear beamformer is similar to those of the linear beamformer presented in Section 5.2. Here, the same idea is applied but, prior to the optimization of the beamformer, a nonlinear transformation is applied in order to map the data in an infinite-dimensional Kernel Reproducing Hilbert Space. Once the data is in this space, it is treated linearly as explained in Chapter 2.

Two main differences with the linear beamformer of this kind of nonlinear ones are important to highlight. First of all, as the function which relates the input and output signals is nonlinear, there is no possibility of drawing a beamshape from the parameters of the beamformer, so all beamshapes will be approximations.

Second, using an adequate nonlinear function, the nonlinear transformation maps the data into an infinite-dimensional space, in which all vectors are linearly independent. Then, the dimension of the subspace in which the transformed signal lies has as many dimensions as data. This gives a more powerful structure to process the signal, having the possibility to deal with more signals than sensors in the antenna.

The input vector $x[n]$ is expressed as in Eq. *(5.1)*

$$x[n] = \mathbf{A}s[n] + \mathbf{g}[n] \tag{5.31}$$

where $\mathbf{A}$ is the matrix of the steering vectors of the incoming signals $s[n]$. The input vector is nonlinearly processed to obtain the output $d[n]$, which is expressed as

$$y[n] = \boldsymbol{w}^T \varphi(\mathbf{x}[n]) = d[n] + \epsilon[n] \tag{5.32}$$

where $d[n]$ is the desired signal at time $[n]$. We assume that a sequence $d[n]$, $n = 1, \cdots, N$ is known and we will use it to train the beamformer.

Note that here we apply a nonlinear transformation $\varphi(x[n])$. As usual, this nonlinear transformation may not be known and the weight vector $\mathbf{w}$ will possibly have an infinite dimension. So, this formulation is in fact not explicit in the structure of the beamformer, but only the equivalent one which uses the dot product $< \varphi(u), \varphi(v) = K(u, v) >$ in the Hilbert space.

Assuming that a set of N observed samples of $x[n]$ are available for training and that the desired signal at the output of the receiver is known, we can directly formulate the dual functional *(4.6)*:

$$L_d = -\frac{1}{2}\psi^H Re\left(\mathbf{K} + \frac{\gamma}{2}\mathbf{I}\right)\psi + Re[\psi^H \mathbf{d}] - (\alpha + \alpha' + \beta + \beta')\mathbf{1}\varepsilon$$

where $\mathbf{d}$ is the vector of all desired signals.

Solving the functional gives the optimum values for parameters ψ_i, and the solution has the same form as that for the linear beamformer except that the dot product between vectors is given by the chosen kernel function $K(\cdot, \cdot)$.

$$d[n] = \sum_{i=1}^{N} \psi_i^* \boldsymbol{x}_i^{*T} \boldsymbol{x}[n] \tag{5.33}$$

Here we do not have an explicit solution for the parameters **w**. Note also that, as we pointed earlier, the Fourier Transform of does not apply here, so we cannot find the beamshape of the beamformer. In other words, where vectors $\mathbf{x} = \{1, \exp(wn) \cdots, \exp(kwn)\}$ are eigen-vectors of linear systems, their transformations $\varphi(\mathbf{x}) = \varphi(\{1, \exp(wn), \cdots, \exp(kwn)\})$ into KRHS are not, in general, so the concept of Fourier Transform disappears on them. The implication of this is that the beamshape will depend on the number and amplitude of arriving sources.

5.5.2   Examples

We compare the performance of the nonlinear (Gaussian kernel) SVM beamformer to those of the linear ones in an environment in which there are N signals, one of them is the desired signals and the other $N-1$ are independent for different number of sensors in the antenna.

In this example, we simulated four directions of arrival. One of them belong to the desired user signal, and the other three are interfering or nondesired independent signals. All signals have unitary amplitude and different randomly chosen phase shifts, and the used modulation is QAM. The channel shows complex-valued additive Gaussian noise. In order to train the beamformers, we use a burst of 100 symbols, and then we test it during 10.000 symbol times, using previously unseen signals. Assuming that the thermal noise power σ_n is known, we choose $\varepsilon = \sigma_n$ and $C = 1000$, though the results with other parameter choice do not show significative differences.

We compare the performance of the nonlinear beamformer with those of the linear one for different values of the signal-to-noise ratio (SNR).

Figure 5.10 shows the results of the simulations. Continuous line graphs correspond to simulations of the nonlinear beamformer, while dashed lines correspond to the simulations of the linear beamformer. The performance of the linear beamformer degrades rapidly when the number of elements is less than 6, as expected. The performance of the SVM beamformer degrades slowly, being able to detect the desired transmitter even when the number of elements is less than the number of angles of arrival.

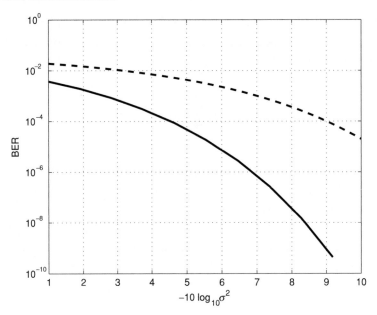

FIGURE 5.10: Simulations of the linear MMSE and the nonlinear SVM with temporal reference beamformers for 7 array elements

5.5.3   Matlab Code

```
function example

%=PARAMETERS=%
n=3;
A=[1 1 1]; %Amplitudes
doa1d=0.5; %DOA 1 Desired
doa2d=0.65; %DOA 1 Desired
doa2=[0.40 0.6 0.7]; p1=0.1;

noise_level=(10:20);
BER=zeros(length(noise_level),2);
noise=10.^(-noise_level/20);

ker='rbf';
C=100;gamma=1e-6;
epsilon=0;
```

```
%=TRAINGING SIGNAL=%
N_train=100; [x_train,B]=generate_data(N_train,n,pi*doa1d);
[x_train]=x_train-0.3*generate_data(N_train,n,pi*doa2d);
for I=1:length(A)
    x_train=x_train+1*A(I)*generate_data(N_train,n,pi*doa2(I));
end
tr_noise=randn(size(x_train))+1j*randn(size(x_train));

%=TESTING SIGNAL=%
N_test=200; [x_test,b]=generate_data(N_test,n,pi*doa1d);
[x_test]=x_test-0.3*generate_data(N_test,n,pi*doa2d);
for I=1:length(A)
    x_test=x_test+1*A(I)*generate_data(N_test,n,pi*doa2(I));
end
tst_noise=randn(size(x_test))+1j*randn(size(x_test));

for j=1:length(noise_level)
    %Adding noise to signals
    x_train_noisy=x_train+noise(j)*tr_noise;
    x_test_noisy=x_test+noise(j)*tst_noise;

    %Normalzation of data and SVM training
    normalization=diag(1./sqrt(sum(x_train_noisy.^2')'/N_train));
    x_tr_nois_norm=(x_train_noisy'*normalization)';
    x_test_nois_norm=(x_test_noisy'*normalization)';

    %Optimization
    [beta, b0,R2] = svr_standard(x_tr_nois_norm,B,p1,C,gamma,
                    epsilon);

    %Test
    R=kernel_matrix(x_test_nois_norm,x_tr_nois_norm,p1);
    output=sign(R*beta+b0); %SVM output
    error=sum(output~=b); %SVM error

    %Representation
    BER(j,:)=[error/(N_test)];
```

```
        semilogy(noise_level,BER)
        xlabel('-\sigma^2_n (AWGN Power)')
        ylabel('Bit Error Rate')
        grid on
end


function [beta, b0,R_] = svr_standard(X,Y,p1,C,gamma,epsilon)

R_=kernel_matrix(X,X,p1); R=[R_ -R_;-R_ R_];
Y=Y(:);
f1=[ones(size(Y')) -ones(size(Y'))];
Y=[Y;-Y];
H=(R+gamma*eye(size(R,1)));
f=ones(size(Y'));
OPTIONS = optimset('LargeScale','off',...
    'diffmaxchange',1e-4,'Diagnostics','off');
alpha=quadprog(H,-Y'+epsilon*f,[],[],f1,0,...
    zeros(size(Y')),C*f,[],OPTIONS);
beta=conj((alpha(1:end/2)-alpha(end/2+1:end)));
b0 = 0;
svi = find( abs(beta) > epsilon);
svii = find(abs(beta) > epsilon & abs(beta) < (C - epsilon));
if length(svii) > 0
    b0  = (1/length(svii))*sum(Y(svii) - ...
        H(svii,svi)*beta(svi).*ones(size(Y(svii))));
end


function [x,b]=generate_data(N,n_elements,phi)
%Simulate data in a N element array
x=zeros(n_elements,N); %Snapshots (row vectors)
b=sign(randn(N,1));      %Data
for i=1:N
    x(:,i)=exp(1j*phi*(0:n_elements-1)')*b(i);
end


function R=kernel_matrix(xtest,x_sv,p1)
N_test=size(xtest,2);% Number of test data
```

```
N_SV=size(x_sv,2);
xtest=xtest';
x_sv=x_sv';
D=buffer(sum(abs([kron(x_sv,ones(N_test,1))...
    - kron(ones(1,N_SV),xtest')']).^2,2),N_test,0);
R=exp(-D/(2*p1^2));% Computation of the kernel matrix
```

5.6   NONLINEAR SVM BEAMFORMER WITH SPATIAL REFERENCE

5.6.1   Structure of the Estimator

In Section 5.3, a linear beamformer with spatial reference has been presented that makes use of the SVM formulation in its linear version. Here we reproduce the generalization of the algorithm to the nonlinear case, presented in [49, 50].

Assume that a nonlinear transformation $\mathbf{x} : \mathbb{R}^N \to \varphi(x) : \mathcal{H}$ from the input space $\mathbb{R}^N$ to a higher dimensional Hilbert space $\mathcal{H}$ whose dot product can be expressed in kernel form, i.e. $\varphi(\mathbf{x}_1)^H \varphi(\mathbf{x}_2) = K(\mathbf{x}_1, \mathbf{x}_2)$. The primal functional for the nonlinear beamformer is the same as in *(5.16)*

$$L_p = \frac{1}{2} \boldsymbol{w}^H \mathbf{R} \boldsymbol{w} + \sum_i L(\xi_i + \xi_i') + \sum_i L(\zeta_i + \zeta_i')$$

but here the autocorrelation matrix $\mathbf{R}$ has the expression

$$\mathbf{R} = \frac{1}{N} \boldsymbol{\Phi} \boldsymbol{\Phi}^H \tag{5.34}$$

$\boldsymbol{\Phi}$ being a matrix whose column i is equal to $\varphi(\mathbf{x}_i)$.

The SVM constraints must be adapted to the nonlinear case as follows. Here, the transformations φ introduce a nonlinear relationship between input and output. Then, we must force the desired signal for each possible direction of arrival $\mathbf{s}_d[n]$ and desired amplitude $r[n]$.

$$\begin{aligned}
\mathbb{R}e\left(r[n] - \boldsymbol{w}^H \varphi(r[n]\mathbf{s}_d[n])\right) &\leq \varepsilon - \xi[n] \\
\mathbb{I}m\left(r[n] - \boldsymbol{w}^H \varphi(r[n]\mathbf{s}_d[n])\right) &\leq \varepsilon - \zeta[n] \\
\mathbb{R}e\left(-r[n] + \boldsymbol{w}^H \varphi(r[n]\mathbf{s}_d[n])\right) &\leq \varepsilon - \xi'[n] \\
\mathbb{I}m\left(-r[n] + \boldsymbol{w}^H \varphi(r[n]\mathbf{s}_d[n])\right) &\leq \varepsilon - \zeta'[n]
\end{aligned} \tag{5.35}$$

The Lagrange analysis gives the result (similar to these of)

$$\boldsymbol{w} = \mathbf{R}^{-1} \boldsymbol{\Phi}_d \psi \tag{5.36}$$

where $\boldsymbol{\Phi}_d = [\varphi(\mathbf{s}_d[1]), \cdots, \varphi(\mathbf{s}_d[N])]$.

The corresponding dual is

$$-\frac{1}{2}\boldsymbol{\psi}^H\left[\boldsymbol{\Phi}_d^H\mathbf{R}^{-1}\boldsymbol{\Phi} + \gamma\mathbf{I}\right]\boldsymbol{\psi} - \mathbb{R}e(\boldsymbol{\psi}^H\mathbf{r}) + \varepsilon\mathbf{1}(\boldsymbol{\alpha} + \boldsymbol{\beta} + \boldsymbol{\alpha}' + \boldsymbol{\beta}') \tag{5.37}$$

where $\mathbf{r} = [r[1], \cdots, r[N]]$.

Expressing the inverse of the autocorrelation matrix as $\mathbf{R}^{-1} = \mathbf{U}\mathbf{D}^{-1}\mathbf{U}^H$, one can rewrite the dual as

$$\frac{1}{2}\boldsymbol{\psi}^H\left[\boldsymbol{\Phi}_d^H\mathbf{U}\mathbf{D}^{-1}\mathbf{U}^H\boldsymbol{\Phi}_d + \gamma\mathbf{I}\right]\boldsymbol{\psi} - \mathbb{R}e(\boldsymbol{\psi}^H\mathbf{r}) + \varepsilon\mathbf{1}(\boldsymbol{\alpha} + \boldsymbol{\beta} + \boldsymbol{\alpha}' + \boldsymbol{\beta}') \tag{5.38}$$

The optimization of the dual *(5.38)* gives us the Lagrange multipliers $\boldsymbol{\psi}$ from which one can compute the optimal weight vector (5.6.1).

We cannot use this algorithm in its present form, because we may not have access to the data into the feature space. In order to make the problem handy, we must use Kernel Principal Component Analysis (KPCA) techniques [53]. Let the autocorrelation matrix in the feature space be defined as in *(5.34)*. The eigenvalues $\mathbf{D}$ and eigenvectors $\mathbf{U}$ of these matrix satisfy $\mathbf{D}\mathbf{U} = \mathbf{R}\mathbf{U}$. Also, the eigenvectors can be expressed as a linear combination of the data set as $\mathbf{U} = \boldsymbol{\Phi}\mathbf{V}$. Combining these two expressions and premultiplying by $\boldsymbol{\Phi}^H$ we get

$$\mathbf{D}\boldsymbol{\Phi}^H\boldsymbol{\Phi}\mathbf{V} = \boldsymbol{\Phi}^H\frac{1}{N}\boldsymbol{\Phi}\boldsymbol{\Phi}^H\boldsymbol{\Phi}\mathbf{V} \tag{5.39}$$

Here, one can find the matrix $\mathbf{K} = \boldsymbol{\Phi}^H\boldsymbol{\Phi}$ whose element i, j is the dot product $\varphi(\mathbf{x}_i)^H\varphi(\mathbf{x}_j) = K(\mathbf{x}_i, \mathbf{x}_j)$, that we can compute.

Putting expression *(5.39)* in *(5.38)* gives the result

$$\frac{1}{2}\boldsymbol{\psi}^H\left[\boldsymbol{\Phi}_d^H\boldsymbol{\Phi}\mathbf{V}\mathbf{D}^{-1}\mathbf{V}^H\boldsymbol{\Phi}^H\boldsymbol{\Phi}_d + \gamma\mathbf{I}\right]\boldsymbol{\psi} - \mathbb{R}e\left(\boldsymbol{\psi}^H\mathbf{r}\right) + \varepsilon\mathbf{1}(\boldsymbol{\alpha} + \boldsymbol{\beta} + \boldsymbol{\alpha}' + \boldsymbol{\beta}') \tag{5.40}$$

The matrix $\mathbf{V}\mathbf{D}^{-1}\mathbf{V}^H = \mathbf{K}^{-1}$ appears, so we can rewrite

$$\frac{1}{2}\boldsymbol{\psi}^H\left[\mathbf{K}_d^H\mathbf{K}^{-1}\mathbf{K}_d + \gamma\mathbf{I}\right]\boldsymbol{\psi} - \mathbb{R}e(\boldsymbol{\psi}^H\mathbf{r}) + \varepsilon\mathbf{1}(\boldsymbol{\alpha} + \boldsymbol{\beta} + \boldsymbol{\alpha}' + \boldsymbol{\beta}') \tag{5.41}$$

The first of the two matrices in the above expression is is the Gram matrix of kernel products whose elements are known. The second is the matrix $\mathbf{K}_d = \boldsymbol{\Phi}^H\boldsymbol{\Phi}_d$ whose elements are $K(\mathbf{x}[n], \mathbf{s}_d[m])$. This dual functional can be now optimized.

Putting Eq. *(5.39)* into (5.6.1) gives the expression of the weights as a function of the dual parameters

$$\mathbf{w} = \mathbf{R}^{-1}\boldsymbol{\Phi}_d\boldsymbol{\psi} = \boldsymbol{\Phi}\mathbf{V}\mathbf{D}^{-1}\mathbf{V}^H\boldsymbol{\Phi}^H\boldsymbol{\Phi}_d\boldsymbol{\psi} = \boldsymbol{\Phi}\mathbf{K}^{-1}\mathbf{K}_d\boldsymbol{\psi} \tag{5.42}$$

and then the support vector machine output for a snapshot $\mathbf{x}[m]$ can be expressed as

$$y[m] = \mathbf{w}^H \varphi(\mathbf{x}[m]) + b = \psi^H \mathbf{K}_d \mathbf{K}^{-1} \Phi^H \varphi(\mathbf{x}[m]) + b = \psi^H \mathbf{K}_d \mathbf{K}^{-1} \mathbf{k}[m] + b \qquad (5.43)$$

where $\mathbf{k}[m] = [K(\mathbf{x}[1], \mathbf{x}[m]), \cdots, K(\mathbf{x}[N], \mathbf{x}[m])]^T$.

For the autocorrelation matrix to be computed in the Hilbert space, the data must be centered, which implies [5]

$$\tilde{\mathbf{K}} = \mathbf{K} - \mathbf{A}\mathbf{K} - \mathbf{K}\mathbf{A} + \mathbf{A}\mathbf{K}\mathbf{A} \qquad (5.44)$$

where $\mathbf{A}$ is an $N \times N$ matrix whose elements are equal to $\frac{1}{M}$, and

$$\tilde{\mathbf{k}}[m] = \mathbf{k}[m] - \mathbf{a}\mathbf{K} - \mathbf{k}[m]\mathbf{A} + \mathbf{a}\mathbf{K}\mathbf{A} \qquad (5.45)$$

where $\mathbf{a}$ is a row vector whose elements are equal to $\frac{1}{N}$.

5.6.2   Examples

We reproduce here the same example as in Subsection 5.5.2 but here we apply the spatial reference SVM (see Fig. 5.11). Continuous line correspond to the nonlinear SVM with spatial reference, and the dashed line corresponds to the linear MVDM. As expected, the performances of both algorithms are similar and clearly outperform the linear ones.

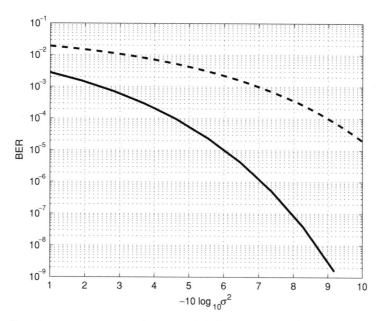

FIGURE 5.11: Simulations of the linear MVDM and the nonlinear SVM with space reference beamformers for 7 array elements

5.6.3   Matlab Code

The following codes are the functions that perform the spatial reference training and test for a BPSK signal. This code can be used together with the code used in Subsection 5.5.2 to reproduce the above experiment.

```
function [beta,b0,R,r,a_d]=SVM_MVDM_TRAIN(x,s_d,ker,par,gamma,
    C,epsilon)
s_d=s_d/norm(s_d(:,1));
x=x/norm(x(:,1));
N=size(x,2);
r=[1;-1]; %Possible Symbols
%These are the steering vectors of the desired directions of
    arrival
a_d=s_d(:,1);
a_d=kron(a_d,r');
%Compute the matrices K and K_d

K=kernel_matrix(ker,x,x,par);
A=ones(size(K))/N;
K_=K-A*K-K*A+A*K*A;
K_d=kernel_matrix(ker,x,a_d,par);

a=ones(size(x,2),size(K_d,2))/N;
K_d_=(K_d-K*a-A*K_d+A*K*a);
%Compute the matrix of the optimization

R_=K_d_'*pinv(K_)*K_d_*N;

1=[ones(size(r'))];
H=(R+gamma*eye(size(R,1)));
f=ones(size(r'));
OPTIONS = optimset('LargeScale','off',...
    'diffmaxchange',1e-4,'Diagnostics','off');
alpha=quadprog(H,-Y'+epsilon*f,[],[],f1,0,...
    zeros(size(Y')),C*f,[],OPTIONS);
beta=conj((alpha(1:end/2)-alpha(end/2+1:end)));
b0 = 0;
```

```
svi = find( abs(beta) > epsilon);
svii = find(abs(beta) > epsilon & abs(beta) < (C - epsilon));
if length(svii) > 0
   b0  = (1/length(svii))*sum(Y(svii) - ...
         H(svii,svi)*beta(svi).*ones(size(Y(svii))));
end

function [y]=SVM_MVDM_TEST(x_sv,s_d,beta,b0,x,ker,par)
s_d=s_d/norm(s_d(:,1));
x=x/norm(x(:,1));
x_sv=x_sv/norm(x_sv(:,1));

N=size(x_sv,2);
r=[1;-1];
%These are the steering vectors of the desired directions of
   arrival
a_d=s_d(:,1);
a_d=kron(a_d,r');

%Compute matrices K and K_d
k=kernel_matrix(ker,x_sv,x,par);
K=kernel_matrix(ker,x_sv,x_sv,par);
A=ones(size(x_sv,2))/size(x_sv,2);
a=ones(size(x_sv,2),size(x,2))/N;
K_=K-K*A-A*K+A*K*A;
k_=k-K*a-A*k+A*K*a;

K_d=kernel_matrix(ker,x_sv,a_d,par);
a=ones(size(x_sv,2),size(a_d,2))/N;
K_d_=(K_d-K*a-A*K_d+A*K*a);
%Compute the matrix
R=K_d_'*pinv(K_)*k_*N;
y=(beta'*R).'+b0;
```

CHAPTER 6

Determination of Angle of Arrival

The capacity of communication systems has limitations due to cochannel interference. For example, in CDMA, users can share the same frequency at the same time, but the number of users is limited by the multiuser interference (MUI) (see, e.g. [54]) due to the fact that the spread spectrum codes are not orthogonal. In real systems, not only the number of users increases the MUI but we also have to deal with the relative distances of them from the receiver. If the receiver antenna is able to determine the angle of arrival (AOA) of each user and then construct different beams for different users and track them, the capacity of the channel can be considerably increased. In TDMA this problem does not exist as users share the spectrum but not the time, but the number of users is limited by the available bandwidth. A truly efficient way to increase the channel capacity is to reuse the available frequencies for those transmitters which are received from different AOAs. This is only possible if the receiver antenna is able to determine the direction of arrival of each receiver and, then, construct a beamshape for each one of the receivers in order to minimize the cochannel interference.

Pseudospectrum computation is used to provide acceptable results in AOA with low computational burden [38,55]. In these methods, the signal is modeled by an AR model. That is, the signal is modeled as the output of a linear AR system for which the input is a white signal. The spectrum of the signal is then equal to the spectrum of the AR. The parameters are fit using a Least Squares algorithm, which is optimal (in the sense of Maximum Likelihood [45]) for Gaussian noise. Nevertheless, the algorithm is highly sensitive to impulse noise and other non-Gaussian noises. Also, the algorithm shows overfitting where the number of samples used to train the system is low.

The most popular algorithms for determining the angle of arrival are the Multiple Signal Classification algorithm (MUSIC), introduced by R. O. Schmidt in 1981, and the ESPRIT algorithm [56]. These algorithms are based on the determination of the noise eigenvectors of the signal and then the computation of its spectrum. Noise eigenvectors are orthogonal to signal eigenvectors. This implies that in the spectrum of noise eigenvectors there will be nulls in those frequencies in which a signal is present [57]. Then, the inverse of the spectrum of noise eigenvectors will show infinite values in the frequencies in which there is a signal. The inverse of the noise spectrums produces a function which shows thin spectral lines in the frequencies of the signals. These algorithms are called hyperresolution algorithms because the thickness of the spectral lines are much smaller that those of the Fourier transform of the signal, so the determination of AOA has much more accuracy.

There are two main disadvantages of these methods. First, they are computationally intensive, so its real-time implementation has a considerable computational burden. The second one is that they assume that the antenna elements are exactly equal, so tolerances in the calibration of them will produce inaccuracies in the AOA computation. If one of the elements fails, the technique also gives inaccurate results.

An alternative to these algorithms is the use of neural networks [58]. Neural networks can be trained offline to determine the AOAs of the signal, so the real-time determination of AOA is less time consuming due to the low computational burden of the operation of the neural network compared to the one of MUSIC or ESPRIT. Second, one can incorporate the calibration tolerances of the antenna element into the signal and then the neural network can compensate them. Then, the ideal element assumption can be removed.

The main drawbacks of classical neural network approaches are that some of them lead to optimization problems which do not have a single solution, and that the complexity of the used structures may produce overfitting . Support Vector Machines are a way to construct linear and nonlinear processors (neural networks) which solve the problem of local minima and overfitting. Here we present three methods for determination of AOA based on linear and nonlinear SVM regressors, and using multiclass SVM classifiers.

The linear method consists of the signal modeling using an AR model and then computing the model DFT. It is shown to have better performance than the classical AR model based AOA using a Least Squares algorithm. The second and third ones have hyperresolution performance with a much lower computational burden.

6.1   LINEAR SVM AOA ESTIMATOR USING REGRESSION

6.1.1   Linear SVM Spectrum Estimator

A direct parametric pseudospectrum estimation can be derived from (4.17). If we set $z_i[n] = x_i[n]$ and use the subarray $\hat{\boldsymbol{x}}_i[n] = x_{i+1}[n] \cdots x_{i+P-1}[n]\}$ then the model is

$$x_i[n] = \sum_{k=1}^{P-1} w_k x_{i+k}[n] + e[n] = \boldsymbol{w}^T \hat{\boldsymbol{x}}_i[n] + e_i[n] \qquad (6.1)$$

This model is just an AR predictor of the sample $x_i[n]$ given the subarray $\hat{\boldsymbol{x}}_i$. The real real-valued counterpart has been introduced in [36].

Assuming white noise, the Z transform of this model with respect the variable i is

$$X(\omega) = \frac{\sigma_e}{1 - \sum_{k=1}^{P-1} w_k e^{jk\omega}} = \frac{\sigma_e}{1 - \boldsymbol{w}^T \boldsymbol{e}} \qquad (6.2)$$

where $\boldsymbol{e} = \{e^{jk\omega}\}$. Applying the result (4.20) to the model weights, the following expression holds

$$X(\omega) = \frac{\sigma_e}{1 - \sum_{i,n} \psi_{i,n} \hat{\boldsymbol{x}}_i[n] \boldsymbol{e}} \qquad (6.3)$$

In the denominator we have the support vectors. We must recall that these vectors are a subset of the whole data set. The quantity of vectors is determined by the value of ε in the cost function *(4.22)*. The selection of this parameter is subject to a trade-off between the quantity of used support vectors to compute the model parameters $\boldsymbol{w}$ and the introduced noise power. If noise $g_i[n]$ is present in the data, then we can write the signal as $x_i[n] = \tilde{x}_i[n] + g_i[n]$, where $\tilde{x}_i[n]$ is the unknown signal (before adding noise) whose spectrum is to be estimated. Then, the expression for the model parameters is

$$\boldsymbol{w} = \sum_{i,n} \psi_{i,n} \tilde{\boldsymbol{x}}_i^*[n] + \sum \psi_{i,n} \boldsymbol{g}_i^*[n] \qquad (6.4)$$

If $\varepsilon = 0$, all the vectors are used, so, in the absence of noise, we will have the best possible estimation of $x_i[n]$. If there is noise in the data, setting $\varepsilon > 0$ will reduce the presence of noise in the parameters. It is intuitively clear that $\varepsilon = 0$ will result in an overfitting of the model, which will be trained to fit to the noisy samples rather than the clean signal. A too much high value of ε will reduce the data, so the estimation error will increase. It can be experimentally seen that for the case of white noise, a good compromise between both limits is setting the value of ε to the noise standard deviation σ_n. This is possible for the case of thermal noise, where its power is usually known. A theoretical justification of this choice for ε can be found in [42].

This method can improve classical LS approach for parametric pseudospectrum computation in two ways. First, the solution is robust against outliers, due to the fact that large errors or outliers are weighted with the same constant C in the solution. Second, when the pseudospectrum technique is used to detect sinusoids buried in noise, it is relatively easy to choose a proper value for ε.

6.1.2   Robustness with Respect to Additive Noise

The Direction of Arrival of two sinusoids of amplitude 1 in white Gaussian noise of power $\sigma^2 = 0.16$ has been estimated using the SVM-AR and the Minimum Variance (MV) methods

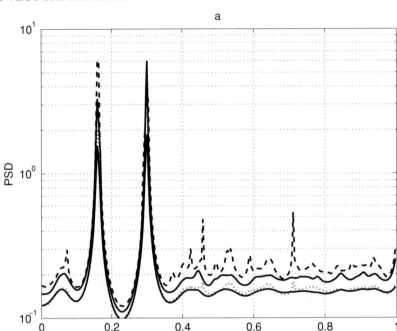

FIGURE 6.1: Mean and variance of SVM pseudospectrum using $\varepsilon = \sigma$ (solid lines) and $\varepsilon = 0$ (dotted and dashed lines). (Source: [10]. Reprinted with permission of Elsevier)

[38]. The number of samples was 30 and $P = 10$ (model order). A value of $C = 0.1$ was set to assure smoothness of the solution. A value of $\delta = 10^{-4}$ was used, but good results were observed for a wide range of this parameter.

Figure 6.1 shows the different performance of SVM-AR as a function of ε. Two hundred trials with $\varepsilon = 0$ and $\varepsilon = \sigma$ were run, and then the mean and the variance of the resulting spectrums were computed. For this experiment, we used a single snapshot ($N = 1$), and subarrays of length $P = 15$ in an array of length $M = 30$. Experiments with $\varepsilon = \sigma$ show much less variance than those with $\varepsilon = 0$. Also, a smaller number and amplitude of artifacts is observed. Figure 6.2 shows the comparison between MV and SVM-AR when $\varepsilon = \sigma$. It can be seen that SVM-AR detects the sinusoids where in the MV the presence of artifacts in the pseudospectrum may be interpreted as signals.

Figure 6.3 shows the mean square error of both methods with respect the noiseless spectrum in a Gaussian white noise environment, while Fig. 6.4 compares both methods for a fixed power of white noise and different powers of impulse (super-Gaussian) noise. The impulse noise is added randomly to a 20% of the samples, and its amplitude has Gaussian density with power σ_I.

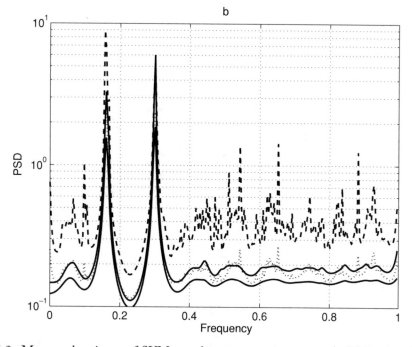

FIGURE 6.2: Mean and variance of SVM pseudospectrum using $\varepsilon = \sigma$ (solid lines) and Minimum Variance method (dotted and dashed lines). (Source: [10]. Reprinted with permission of Elsevier)

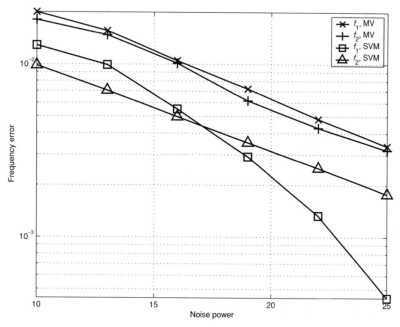

FIGURE 6.3: DOA error of the SVM and the MV spectrums in additive Gaussian white noise for a signal containing two DOAS

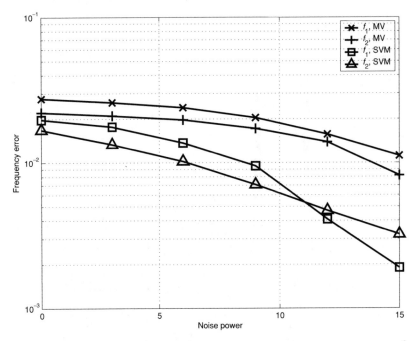

FIGURE 6.4: DOA error of the SVM and the MV spectrums in AWGN of power 10^{-3} plus impulse noise for a signal containing two DOAS

6.1.3   Matlab Code

```
function example

%=PARAMETERS=%
n=15;
A=[1 1 1]; %Amplitudes
doa1d=0.5; %DOA 1 Desired
doa2d=0.65; %DOA 1 Desired
doa2=[0.40 0.6 0.7]; p1=0.1;

noise_level=10;
noise=10.^(-noise_level/20);
C=100;gamma=1e-6;
epsilon=0;

%=TRAINGING SIGNAL=%
N_train=100; [x_train,B]=generate_data(N_train,n,pi*doa1d);
```

```
[x_train]=x_train-0.3*generate_data(N_train,n,pi*doa2d);
for I=1:length(A)
    x_train=x_train+1*A(I)*generate_data(N_train,n,pi*doa2(I));
end
tr_noise=randn(size(x_train))+1j*randn(size(x_train));

%Adding noise to signals
x_train_noisy=x_train+noise*tr_noise;

%Normalzation of data and SVM training
normalization=diag(1./sqrt(sum(x_train_noisy.^2')'/N_train));
x_tr_nois_norm=(x_train_noisy'*normalization)';

%Optimization
[beta, b0,R2] = svr_standard(x_tr_nois_norm,B,p1,C,gamma,epsilon);
w=beta'*x_tr_nois_norm';

spectrum=abs(fftshift(fft(w,256)));
semilogy(spectrum)

function [beta, b0,R_] = svr_standard(X,Y,p1,C,gamma,epsilon)
R_=kernel_matrix(X,X); R=[R_ -R_;-R_ R_];
Y=Y(:);
f1=[ones(size(Y')) -ones(size(Y'))];
Y=[Y;-Y];
H=(R+gamma*eye(size(R,1)));
f=ones(size(Y'));
OPTIONS = optimset('LargeScale','off',...
    'diffmaxchange',1e-4,'Diagnostics','off');
alpha=quadprog(H,-Y'+epsilon*f,[],[],f1,0,...
    zeros(size(Y')),C*f,[],OPTIONS);
beta=conj((alpha(1:end/2)-alpha(end/2+1:end)));
b0 = 0;
svi = find( abs(beta) > epsilon);
svii = find(abs(beta) > epsilon & abs(beta) < (C - epsilon));
if length(svii) > 0
```

```
    b0  = (1/length(svii))*sum(Y(svii) - ...
         H(svii,svi)*beta(svi).*ones(size(Y(svii)))));
end


function [x,b]=generate_data(N,n_elements,phi)
%Simulate data in a N element array
x=zeros(n_elements,N); %Snapshots (row vectors)
b=sign(randn(N,1));    %Data
for i=1:N
     x(:,i)=exp(1j*phi*(0:n_elements-1)')*b(i);
end


function R=kernel_matrix(xtest,x_sv)
R=xtest'*x_sv;
```

6.2   NONLINEAR AOA ESTIMATORS

6.2.1   Structure of an AOA Problem on Nonlinear Regression

In [59], Pastorino et al. showed that a nonlinear regressor can be used to solve the problem of AOA of several signals. Basically, the idea behind it is to use an SVR to approximate the unknown function χ that relates the received signals and their angles of arrival.

A number of input/output pairs are created by considering an N fixed number of angles $(\theta_0, \theta_1, \ldots, \theta_{N-1})$ and their corresponding normalized array input vectors ($\mathbf{z}_0, \mathbf{z}_1, \ldots, \mathbf{z}_{N-1}$). These input/output pairs can be used to evaluate the function $\widetilde{\chi}$ which approximates the unknown desired function χ, i.e.,

$$\widetilde{\chi}(\mathbf{z}) = \langle \boldsymbol{w}, \varphi(\mathbf{z}) \rangle + b \tag{6.5}$$

Here Φ is the nonlinear function that is used for transforming the input array data from its original space to a higher dimensional space. The regression risk associated with this SVR is given by (see Appendix A)

$$L_m = C \sum_{i=0}^{N-1} \ell(\mathbf{z}^i, \hat{\theta}_i) + \frac{1}{2} \|\boldsymbol{w}\| \tag{6.6}$$

where $C =$ constant and $\ell(\mathbf{z}^i, \hat{\theta}_i)$ is the ε-insensitive cost function that was described in Chapters 2 and 3. More specifically, this function can be expressed as (see Eq. *(2.14)*)

$$\ell(\mathbf{z}^i, \hat{\theta}_i) = \begin{cases} 0 & |\hat{\theta}_i - \widetilde{\chi}(\mathbf{z}^i)| < \varepsilon \\ |\hat{\theta}_i - \widetilde{\chi}(\mathbf{z}^i)| - \varepsilon & |\hat{\theta}_i - \widetilde{\chi}(\mathbf{z}^i)| > \varepsilon \end{cases} \tag{6.7}$$

with ε being the allowable error during training.

Using the dual optimization problem (see Eq. *(3.15)*), $\mathbf{w}$ can be written in terms of the input data $\mathbf{z}$ as [59]

$$\boldsymbol{w} = \sum_{i=0}^{N-1} (\alpha_i - \alpha_i')\,\varphi(\boldsymbol{z}^i) \tag{6.8}$$

where α_i and α_i' are unknown coefficients. Substituting (6.8) into (6.5) yields

$$\tilde{\chi}(\boldsymbol{z}) = \sum_{i=0}^{N-1} (\alpha_i - \alpha_i')\langle\varphi(\boldsymbol{z}^i), \varphi(\boldsymbol{z})\rangle + \boldsymbol{b} \tag{6.9}$$

or

$$\tilde{\chi}(\boldsymbol{z}) = \sum_{i=0}^{N-1} (\alpha_i - \alpha_i')K(\boldsymbol{z}^i, \boldsymbol{z}) + \boldsymbol{b} \tag{6.10}$$

In (6.9) the kernel function that is used on the original space is given by (see Eq. (3.6)):

$$K(\boldsymbol{z}^i, \boldsymbol{z}) = e^{-0.5\left\| \boldsymbol{z} - \boldsymbol{z}^i \right\|^2} \tag{6.11}$$

The parameter C in this case was set equal to 0.82. This number was determined a trade-off between the capability of the SVR algorithm to estimate the angles of arrival from signals in the training set and the angles of arrival of the signals that were not included in the training set.

Following the procedure outline in Subsection 2.2.3, in order to solve the constrained optimization problem in (6.10), we must use Lagrange multipliers to convert it into an unconstrained problem. Thus, we get [59]

$$L(\alpha, \alpha') = -\epsilon \sum_{i=0}^{N-1} (\alpha_i + \alpha_i') + \sum_{i=0}^{N-1} \hat{\theta}^i(\alpha_i' - \alpha_i) - \frac{1}{2} \sum_{i,j=0}^{N-1} (\alpha_i' + \alpha_i)(\alpha_i' - \alpha_i)K(\boldsymbol{z}^i, \boldsymbol{z}^j) \tag{6.12}$$

To demonstrate the validity of this approach, Pastorino et al. [59] examined several scenarios. Figure 6.5 shows a case where there are two impinging signals, separated by 15°. The figure shows how well the SVR can estimate the angle of arrivals for the two signals as a function of sample number. The number of sample used for the test set was 166. Figure 6.6 depicts another example where the signals are separated by 17.5°. In Fig. 6.7 the SVR algorithm is compared with two other techniques, MUSIC and Neural Networks.

Next, another approach, using multiclass classification, for determining the angles of arrival is presented.

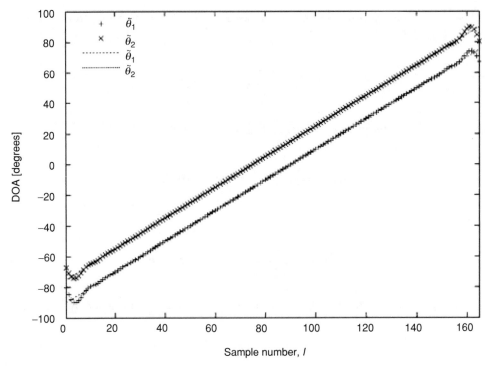

FIGURE 6.5: Comparison of actual and estimated values for AOAs for two impinging signals (separated by 15°) versus the sample number. (Source: [59]. Reprinted by permission of IEEE)

6.3 NONLINEAR SVM ESTIMATOR USING MULTICLASS CLASSIFICATION

6.3.1 Structure of an AOA Based on a Multiclass Classifier

Rowher et al. [60, 61] developed a multiclass Least Squares-SVM (LS-SVM) architecture for AOA estimation as applied to a CDMA cellular system. One-vs-one multiclass classification is based on the extension of binary LS-SVMs. For M distinct classes one can construct $M(M - 1)/2$ hyperplanes that can separate these classes with a maximum margin during the LS-SVM training phase. For one-vs-one multiclass classification the Decision Directed Acyclic Graph (DDAG) technique [62] is employed. This technique uses a tree structure to compare the test data to each of the hyperplanes. Through a process of elimination the closest label is assigned to the input vector data. The LS-SVM algorithm developed in [60] for AOA estimation is based on the DDAG architecture. Figure 6.8 shows an example of a DDAG architecture with each each node containing a binary LS-SVM classifier of the ith and jth classes.

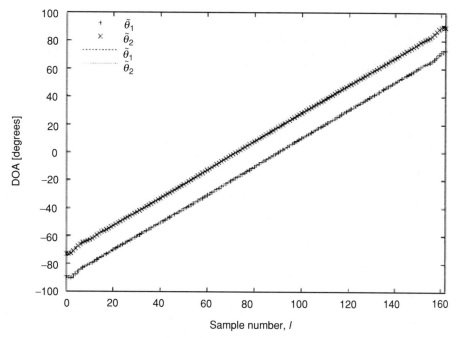

FIGURE 6.6: Comparison of actual and estimated values for AOAs for two impinging signals (separated by 17.5°) versus the sample number. (Source: [59]. Reprinted by permission of IEEE)

This LS-SVM AOA estimation algorithm can be broken in the following 5 stages:

- **Preprocessing for SVM Training Data**
  - Generate the $L \times N$ training vectors for the M SVM classes. L represents the number of antenna elements in the array and N the number of input data samples used for training.
  - Generate the sample covariance matrices from the data vector.
  - Calculate the signal eigenvector from each of the M sample covariance matrices.
  - Calculate the L projection vectors for each of the M classes.
  - Store the projection vectors for the training phase and the eigenvectors for the testing phase.
- **LS-SVM Training**
  - Using the projection vectors train the $M(M-1)/2$ nodes with the one-vs-one LS-SVM algorithm.
  - Store the LS-SVM Lagrange multipliers α_i and the bias term b (see Eq. *(2.1)*).

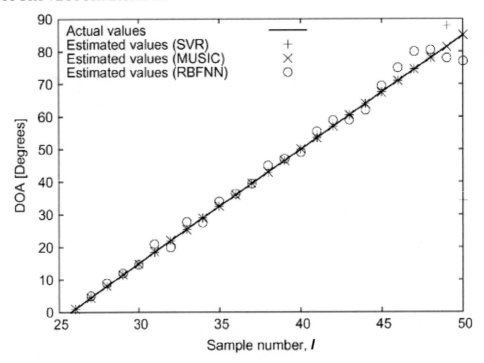

FIGURE 6.7: Comparison of estimated values for AOAs versus the sample number against the standard Music algorithm and with a NN-based technique.(Source: [59]. Reprinted by permission of IEEE)

- **Preprocessing for SVM Testing**
  - Determine the $L \times N$ input signal from the antenna array.
  - Obtain the sample covariance matrix with samples from the $L \times N$ data vector.
  - Find the eigenvectors for the signal subspace and the noise subspace.
  - Calculate the covariance matrices for each eigenvector.

- **LS-SVM Testing for the ith/jth DDAG Node**
  - Obtain two projection vectors with the desired eigenvector covariance matrix and the ith and jth eigenvectors stoted during the training phase.
  - Test both projection vectors against the LS-SVM hyperplane for the ith/jth node.
  - Evaluate the mean value of the two LS-SVM output vectors (labels). Select the closest mean value to a decision boundary (0 or 1).
  - Compare this value to the label definition at the DDAG node, then assign the proper output label.
  - Repeat the same process for the next DDAG node in the evaluation path or declare the final DOA label.

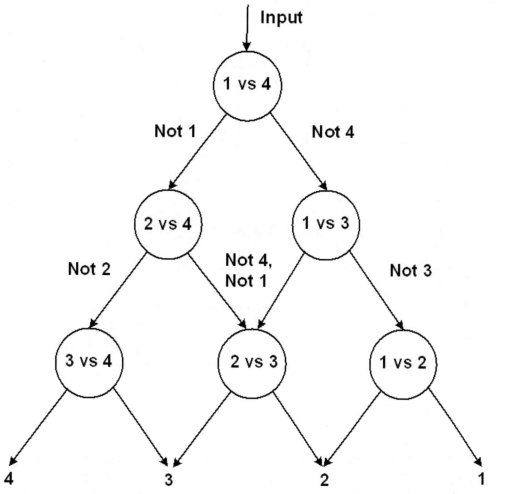

FIGURE 6.8: LS-SVM for DOA estimation for five classes with 4 degrees of separation between each

- **LS-SVM DDAG Error Control**
    - Review the mean square error calculations for the DDAG evaluation path.
    - Apply error control and classify the label as either an accurate AOA estimate or as noise.

6.3.2   Simulation Results

Simulations of the LS-SVM DDAG AOA estimation algorithm are based on a complex system model that includes amplitude and phase distributions that describe the communication channel in use.

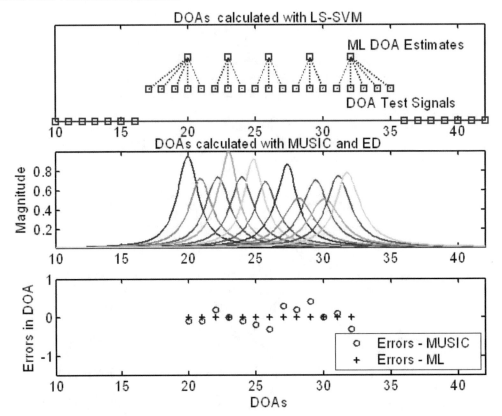

FIGURE 6.9: LS-SVM for DOA estimation, four classes with 10 degrees separation between each

To demonstrate the functionality of this approach an antenna array with eight elements was used The LS-SVM system includes five AOA classes and six DDAG nodes. Figure 6.9 depicts results for a different degree range per class. To test the DDAG capabilities the simulations were automated to test a wide range of AOAs. As can been seen from Fig. 6.9 the LS-SVM DDAG AOA estimation algorithm is extremely accurate. Additional simulations show that the LS-SVM DDAG system accurately classifies the AOAs for three to ten classes and AOA ranges from 1 degree to 20 degrees.

CHAPTER 7

Other Applications in Electromagnetics

SVMs have also been used in several other applications in electromagnetics such as automatic target recognition in radar [63, 64], remote sensing [65, 66] and in the intelligent alignment of waveguide filters. Moreover, SVMs have been used in the fault detection of digital microwave radio communication equipment and electrical machines [67]. Here we are going to show a couple of recent examples that have been used successfully in electromagnetics.

7.1 BURIED OBJECT DETECTION

SVMs are very suitable for real-time inverse scattering problems in electromagnetics since the inverse scattering problem can be casted as a constrained quadratic optimization problem. Bermani et al. in [68] considered a two-dimensional space problem shown in Fig. 7.1

The figure represents a pipe buried in a lossy inhomogeneous medium. The parameter characterizing the scatterer are ϵ_r and σ. if the line source located at (x_t, y_t) is radiating then the scattered field at the observation point (x_r, y_r) can be expressed as:

$$E^{scat}(x_r, y_r) = E_o^{sc}(x_r, y_r) + k^2 \int_D E_o(x, y) \cdot G(x_r - x, y_r - y) \, F$$
$$\times [(x - x_o, y - y_o), \rho, \epsilon_r, \sigma] \, dx \, dy \qquad (7.1)$$

where E_o^{sc} is the electric field inside the domain filled with the background lossy medium. G is the Green's function [68] and $F(x, y)$ is the dielectric profile of the medium. The aim is to determine the location, shape, and dielectric properties of the buried object from the scattered field $E^{scat}(x_r, y_r)$. Mathematically, the problem can be casted as

$$\underline{\gamma} = \Phi[E^{scat}] \qquad (7.2)$$

where $\underline{\gamma}$ is the scatterer data array described in terms of position, radius, ϵ_r and σ. E^{scat} can also be written as an array of sample data as shown in Fig. 7.1.

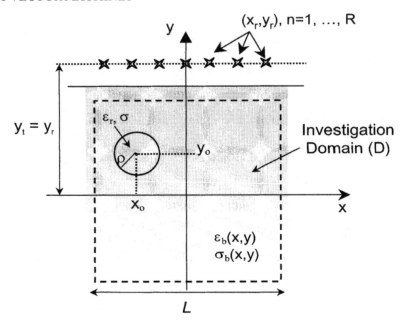

FIGURE 7.1: Geometry of a buried object. (Source: [68]. Reprinted by permission of IEEE)

To use SVM on Eq. (7.2) the problem must be reformulated as a *regression* problem following the procedure outlined in Chapter 2. The unknown function Φ must be evaluated from several sets of I/O data pairs of vectors $\{\gamma_n, E_n^{escat}\}$ for $n = 1, \ldots, N$.

In [68], Bermani et al. verified the SVM algorithm using a buried circular cylinder with an electric line source, located at $\lambda_o/6$ above the air–soil interface. The dielectric constant of the cylinder is $\epsilon_r = 5.0$ and its radius $\rho = \lambda_o/12$. The subsurface region is characterized by $\epsilon_b = 20$ and $\sigma_b = 10^{-2}$ S/m. The E^{scat} data was generated in this example using an FEM code but any code or experimental data can be used. 16 equally spaced points were used to sample the field at $y_r = \frac{2}{3}\lambda_o$. Figure 7.2 depicts the estimated versus the actual scatterer properties. Some additional work in using SVMs for the detection of buried objects can be found in [69].

7.2   SIDELOBE CONTROL

In Chapters 5 and 6 it was shown SVMs can be used for beamforming and estimating the direction of arrival (DOA). Gaudes et al. [70] demonstrated how sidelobe control with beamforming can be optimized using SVMs. This was achieved by modifying the basic MVDR beamforming problem by adding additional constraints as a regularization term of the array output power and the problem was converted into a quadratic programming problem similar to that of a classical Support Vector Regressor. The derivation of the optimization problem can be found in [70]. Here we show some of the results obtained from that work.

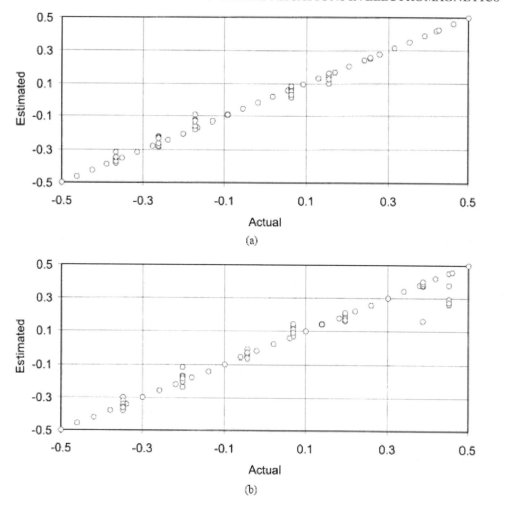

FIGURE 7.2: Estimated versus real scatterer properties object. (a) x_o/λ_o and (b) y_o/λ_o. (Source: [68]. Reprinted by permission of IEEE)

7.2.1 Example 1

The first example assumes that the source steering vector is known exactly and that power of the signal of interest (SOI) has a power level of 10 db and the interfering signals are at the 30 db level. Moreover, the angle of arrival for the SOI is at $\theta_s = 0°$ and the directions of arrival for the interfering signals are at $\theta_1 = -30°$, $\theta_2 = 30°$, and $\theta_3 = 70°$. Fifty snapshots ($N = 50$) were used to compute the correlation matrix. Figure 7.3 depicts the array patterns of the SVM approach for two cases of SNR as compared to two other beamformers (the Capon Beamformer and (SpheRCB) a robust array beamformer that includes a spherical type of constraint [71]). The vertical lines in the figure are used to indicate the location of both the SOI and the interfering

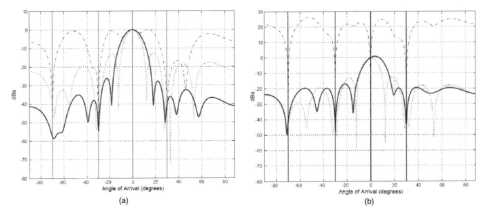

FIGURE 7.3: Patterns for 3 different beamformers: SVM (solid line), SpherCB (dotted line), and the Capon Beamformer (dash-dotted line). (a) SNR = 5 dB, (b) SNR = 30 dB. (Source: [70]. Reprinted by permission of IEEE)

signals. One can see from the this figure that the SVM approach yields lower sidelobes, especially for low SNRs.

7.2.2   Example 2

In this case the steering vector included some error. The actual desired DOA was set at $\theta_a = 0°$ and assumed desired signal (that is the error) at $\theta_s = 2°$. The rest of the parameters are the same as in the previous example. From Fig. 7.4 one can see that the Capon beamformer classifies the signal at $2°$ as an interfering signal. The SVM and SpheRCB approaches classify the signal correctly but only the SVM yields lower sidelobe levels as well.

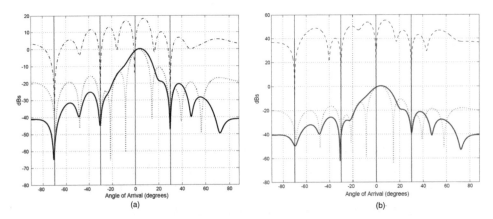

FIGURE 7.4: Patterns for 3 different beamformers: SVM (solid line), SpherCB (dotted line), and the Capon Beamformer (dash-dotted line). (a) SNR= 5 dB, (b) SNR= 30 dB. (Source: [70]. Reprinted by permission of IEEE)

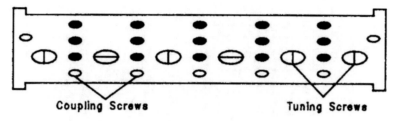

Coupling Screws   Tuning Screws

FIGURE 7.5: Top view of a six-cavity waveguide filter. (Source: [72]. Reprinted by permission of IEEE)

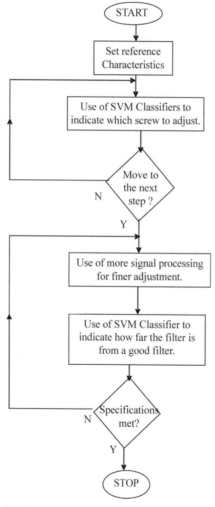

FIGURE 7.6: Flow chart of the process used

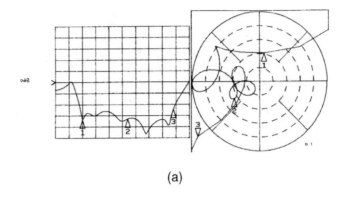

(a)

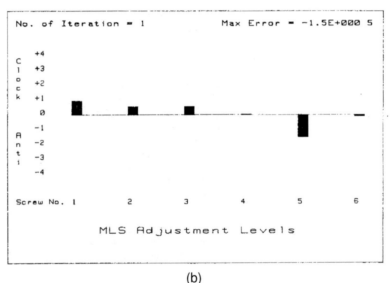

(b)

FIGURE 7.7: (a) The S_{11} of an untuned filter. Center frequency = 10.531 GHz, bandwidth = 0.136 GHz, and minimum loss = −26 db. (b) The MLS output of the filter. (Source: [72]. Reprinted by permission of IEEE)

7.3   INTELLIGENT ALIGNMENT OF WAVEGUIDE FILTERS

In 1989, Mirzai et al. [72] adopted a machine learning system in a way that it can assist an unskilled operator to accurately tune waveguide filters. Figure 7.5 shows the top view of a cavity waveguide filter with two types of screws. the bigger size screw is used for tuning purposes, whereas the other for coupling purposes.

The process of applying the SVM approach in this case is shown in the flow chart in Fig. 7.5. As shown it is a two-step process that can be used to tune waveguide filters in a very systematic and accurate way. The high level of interaction between adjacent screws and their

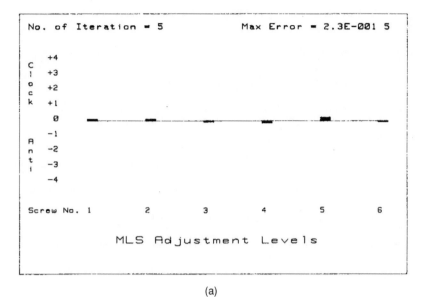

(a)

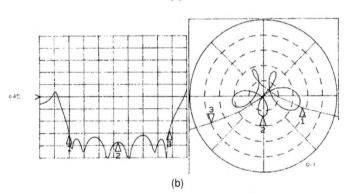

(b)

FIGURE 7.8: (a) The MLS output of the filter after four adjustments. (b) The S_{11} of the tuned filter. Center frequency = 10.532 GHz, bandwidth = 0.138 GHz, and minimum loss = −32 db. (Source: [72]. Reprinted by permission of IEEE)

effect on the sensitivity of the filter is also an issue that has to be taken care of during the training of the SVM.

Figures 7.6 and 7.7 show how the output of the MLS (machine learning) algorithm and the S_{11} characteristics for an untuned and a tuned filter, respectively.

APPENDIX A

The Principle of Structural Risk Minimization

A.1 ACTUAL RISK AND EMPIRICAL RISK

The underlying task in classification or regression is to find a rule or function which assigns a scalar y to a vector x. One possibility consists of choosing a family of parametric functions $f(x, w)$, where w are the parameters. Then, the function $f(x, w) : \mathcal{X} \to \mathcal{Y}$ is estimated by adjusting w.

Adjustment of parameters has to be done accordingly to a given optimality criteria. The actual performance of the machines can be measured by its risk

$$L(f) = \int \ell(f(w, x), y) d P(x, y) \tag{A.1}$$

where $P(x, y)$ is the probability distribution of the pairs (x, y) and $\ell(\cdot)$ is a function chosen to take a measure of the output error, or *loss function*. In binary classification, the indicatrix function $\mathbf{I}(f(x, w = y))$ is often used. The most commonly used loss function for regression are the square function $\ell = (f(w, x) - y)^2$ and the absolute value $\ell = |f(x, w) - y|$.

The measure $L(f)$ is the so-called risk or generalization error. As $P(x, y)$ is not known, all knowledge must be taken from the available data set $\{x, y\}$.

The first approach one can take is to minimize the measured empirical risk by taking a suitable loss function.

$$L_m = \frac{1}{N} \sum_{i=1}^{N} \ell(f(x, w), y) \tag{A.2}$$

where N is the number of available samples. The intrinsic principle applied to this procedure is the assumption of that the risk L of the machine monotonically approaches the empirical risk L_m. In general, this assumption is not true. The empirical risk of the function $(f(w, x), y))$ can be arbitrarily low depending on the complexity of it.

For example, one may attempt to minimize the regression function for the system $y_i = Ax_i + g_i$ where g_i is the observation noise. If the family of Gaussian functions $f(x, \boldsymbol{w}, \sigma) = \sum_k w_k G(x, \sigma)$ is chosen as a regressor, the parameter set $\boldsymbol{w}$ which reduces to zero the empirical risk can be obtained. But the resulting function will not approximate well the linear function $y = Ax$ unless the noise is zero.

This is an example of the phenomenon known as *overfitting*. Overfitting is a consequence of a lack of knowledge of the statistical properties of the data. The first approach of minimizing the empirical risk assumes that overfitting will monotonically decreases with the number of data, but this may not be true, as in the example above.

To make the empirical risk monotonically approach the actual risk, one classical strategy is to use a sequential training algorithm, and use a second data set called *validation set*, not used for training, to measure its empirical risk. The goal is to minimize the empirical risk of the validation set.

Another strategy which one can use is to use a technique to control the complexity of the machine [73]. It is clear that if we constrain the expression ability of the above machine, we will obtain a smoother solution, which will approximate better the given system. Trying to control the complexity of a machine is often called *regularization* [27].

A.2   COMPLEXITY CONTROL BY MEANS OF SRM

The regularization technique used in Support Vector machines is the so-called Structural Minimization Principle. It has been proven [1,3] that the actual risk $L(f)$ of a machine is bounded with probability $1 - p$ by

$$L_m + \sqrt{\frac{h \log(2N/h) - \log(p/4)}{N}} \qquad (A.3)$$

where h is called Vapnik–Chervonenkis (VC) dimension. The VC dimension is a measure of the complexity of the machine. The second term of the expression is the *Structural Risk* of the machine. Two terms control the generalization ability of the machine, the first one being the number of samples N, as is intuitive. The other one is the VC dimension. By decreasing this parameter (equivalently decreasing the complexity) the generalization ability increases. On the other side, if the VC dimension decreases too much, the resulting machine may not be able to express the data, thus increasing the empirical risk.

Vapnik also provides a proof of that for a linear machine

$$y = \boldsymbol{w}^T \boldsymbol{x} + b \qquad (A.4)$$

used to approximate the set x_i, y_i with $||w|| < A$ and being the the VC dimension is bounded by

$$h \leq A^2 R^2 \qquad\qquad (A.5)$$

R^2 being the smallest ball containing the data.

This gives us a rule to minimize the complexity: minimizing the norm of the vector w maximizes the generalization ability. One can also see from this result that the complexity of a linear machine not just consist of its number of parameters in w, but also of its norm.

There is no mean to measure the VC dimension of an arbitrary machine, but the above rule can still be used to minimize the complexity of certain nonlinear machines. These machines are constructed using the rather old kernel trick, which are explained in Chapter 3.

APPENDIX B

Linear Support Vector Classifier

B.1   MINIMIZATION OF THE SRM IN A LINEAR CLASSIFIER

A binary classifier is a function $f : \mathcal{X} \to \mathcal{Y}$ which maps a set of vectors x into scalars (or labels) $y \in +1, -1$. A linear classifier is constructed by means of a separating hyperplane

$$\Pi \equiv w^T x + b = 0 \qquad (B.1)$$

The classification rule can be written as

$$y = sign(w^T x + b) \qquad (B.2)$$

The Structural Minimization Principle here is applied by controlling the norm of the vector w. An intuitive explanation of this is choosing the following heuristic: provided all the available sample can be classified without any missclassification, the best one can do is to place the separating hyperplane at the maximum possible distance of the nearest points x_i subject to the constrain (Fig. B.1)

$$y_i(w^T x_i + b) \geq 1 \qquad (B.3)$$

This constraint means, simply, that the sign of the classifier output and the sign of the label must be equal in all cases.

In practical situations, the samples are not linearly separable, so constraints (B.3) cannot be satisfied. For that reason, slack variables must be introduced to account for the nonseparable samples. Then, the optimization criterium consist of minimizing the (primal) functional [74].

$$L_p = \frac{1}{2}||w||^2 + C\sum_{i=1}^{N} \xi_i \qquad (B.4)$$

subject to the constraints

$$y_i(w^T x_i + b) \geq 1 - \xi_i \qquad (B.5)$$

The values of ξ_i must be constrained to be nonnegative. If the sample x_i is correctly classified by the hyperplane and it is out of the margin, its corresponding slack variable is $\xi_i = 0$. If it is well classified but it is into the margin, $0 < \xi_i < 1$. If the sample is missclassified, then $\xi_i > 1$.

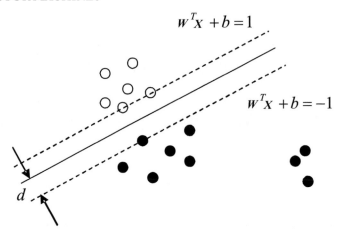

FIGURE B.1: A rule to optimize a separating hyperplane is to maximize its distance to the nearest samples

   The term $1/2$ in the functional (B.4) is placed for computational convenience. The term C is a parameter to adjust the trade-off between the minimization of the parameters (minimization of the VC dimension) and the minimization of the empirical error. A low value of C will lead to a low complex machine, but too low values will not be able to express the data, while a too much high value of C will lead to a machine whose empirical error is low, but that will have poor generalization ability. The value of C is, thus, a parameter to heuristically adjust, though several approaches avoid the search for C in some cases [43].

B.2   SOLVING THE FUNCTIONAL

In order to optimize the constrained functional (B.4), we need to make use of Lagrange multipliers. First, the constraints must be rewritten as

$$y_i(\boldsymbol{w}^T\boldsymbol{x}_i + b) - 1 + \xi_i \geq 0$$
$$\xi_i \geq 0 \tag{B.6}$$

We must minimize a functional with respect to $\boldsymbol{w}$ and ξ_i and we need to maximize it with respect to the constraints. The corresponding primal-dual functional to minimize is then constructed by substracting a linear combination of the constraints to the primal

$$L_{pd} = \frac{1}{2}||\boldsymbol{w}||^2 + C\sum_{i=1}^{N}\xi_i - \sum\alpha_i[y_i(\boldsymbol{w}^T\boldsymbol{x}_i + b) - 1 + \xi_i] - \sum_{i=1}^{N}\mu_i\xi_i \tag{B.7}$$

where α_i, β_i are the Lagrange multipliers. The Karush–Kuhn–Tucker (KKT) conditions [75] for this problem are

$$\frac{\partial L_{pd}}{\partial \boldsymbol{w}} = 0$$

$$\frac{d L_{pd}}{d b} = 0 \tag{B.8}$$

$$\frac{\partial L_{pd}}{\partial \xi_i} = 0$$

$$\alpha_i, \mu_i \geq 0 \tag{B.9}$$

$$\alpha_i y_i[(\boldsymbol{w}^T \boldsymbol{x}_i + b) - 1 + \xi_i] = 0 \tag{B.10}$$

$$\mu_i \xi_i = 0 \tag{B.11}$$

Applying the KKT conditions (B.8) leads to

$$\frac{\partial L_{pd}}{\partial \boldsymbol{w}} = \boldsymbol{w} - \sum_{i=1}^{N} \alpha_i y_i x_i = 0 \tag{B.12}$$

$$\frac{d L_{pd}}{d b} = - \sum_{i=1}^{N} \alpha_i y_i = 0 \tag{B.13}$$

$$\frac{\partial L_{pd}}{\partial \xi_i} = C - \alpha_i - \mu_i = 0 \tag{B.14}$$

The result (B.12) gives the solution for the parameters

$$\boldsymbol{w} = \sum_{i=1}^{N} \alpha_i y_i x_i \tag{B.15}$$

Combining these equations with the primal-dual leads to the dual functional

$$L_d = -\frac{1}{2} \sum_{i=1}^{N} \sum_{j=1}^{N} \alpha_i \alpha_j y_i y_j x_i{}^T x_j + \sum_{i=1}^{N} \alpha_i \tag{B.16}$$

which can be written in matrix format as

$$L_d = -\frac{1}{2} \boldsymbol{\alpha}^T \boldsymbol{Y} \boldsymbol{R} \boldsymbol{Y} \boldsymbol{\alpha} + \boldsymbol{\alpha} \tag{B.17}$$

$\boldsymbol{\alpha}$ being a column vector containing all the Lagrange multipliers α_i, $\boldsymbol{Y}$ a diagonal matrix of the form $\boldsymbol{Y}_{ii} = y_i$, and $\boldsymbol{R}$ being the matrix of dot products

$$\boldsymbol{R}_{ij} = \boldsymbol{x}_i^T \boldsymbol{x}_j \qquad (B.18)$$

Note that condition (B.11) eliminates the last term of the functional (B.7). Also, it is easy to see from condition (B.11) and (B.12) that

$$
\begin{aligned}
&0 \le \alpha_i \le C \\
&\text{if } \alpha_i < C, \text{ then } \xi_i = 0 \qquad\qquad (B.19) \\
&\text{if } \xi_i > 0, \text{ then } \alpha_i = C
\end{aligned}
$$

Then, the slack variables does not appear in the dual formulation because term $C \sum \xi_i$ cancels with $\sum \alpha_i \xi_i$.

This functional is a quadratic form, and it can be solved using quadratic programming [32] or other techniques [76].

Only a subset of the Lagrange multipliers will have a nonzero value in the solution, while the others will vanish. Their associated samples are the so-called Support Vectors.

In order to compute the value of b one has to take into account that if $\alpha_i < C$, then $\xi_i = 0$, as stated in (B.19). Then, it is easy to compute b from condition (B.10) for any sample $\boldsymbol{x}_i$ for which $\alpha_i < C$. In practice, it is numerically convenient to average the result of b for all samples with $\alpha_i < C$.

Some of the eigenvectors of $\boldsymbol{R}$ may have very small associate eigenvalues, leading to an ill-conditioned problem. This inconvenience is overcome as usual by adding a small identity matrix. Then, the dual to solve will be

$$L_d = -\frac{1}{2}\boldsymbol{\alpha}^T \boldsymbol{Y}[\boldsymbol{R} + \gamma \boldsymbol{I}]\boldsymbol{Y}\boldsymbol{\alpha} + \boldsymbol{\alpha} \qquad (B.20)$$

The introduction of this identity matrix is known to have an equivalent effect as modifying the loss measure over the slacks in (B.4). There, a linear loss function is applied. Adding the identity is equivalent to apply a quadratic cost function to the slacks whose value is less than γC and a linear cost function to the others. We formalize this in Appendix C for the case of regression SVM.

APPENDIX C

Linear Support Vector Regressor

C.1 APPLICATION OF THE SRM TO REGRESSION

A regressor is a function $f : \mathcal{R}^n \to \mathcal{R}$ which maps a set of vectors $\boldsymbol{x} \in \mathcal{R}^n$ into a set $y \in \mathcal{R}$. A regressor is constructed as

$$y(\boldsymbol{x}) = \boldsymbol{w}^T \boldsymbol{x} + b + \epsilon(\boldsymbol{x}) \qquad (C.1)$$

where $\epsilon(\boldsymbol{x})$ accounts for the approximation error. Regression has been widely used in many areas, among them, model estimation, linear prediction [55], equalization [45], others, and, of course, array signal processing [57], in time and frequency domains.

Classical approaches use to apply a Minimum Squared Error strategy (MSE) of the available data, making the assumption that it is corrupted by additive Gaussian white noise (AWGN). Then, the best one can do is to use a MSE algorithm, as it is the Maximum Likelyhood (ML) function for AWGN. Nevertheless, one cannot directly minimize the actual Least Squares function because the parameters of the noise distribution are unknown. A simple MSE may then result in overfitting. In order to make the resulting machine robust against overfitting, a number of regularization techniques exist. Moreover, MSE is not the best one can do when, in addition to the unavoidable observation AWGN, other non-Gaussian noises are present. This is often the case in communications. In these applications multiuser noise, which is not Gaussian [77] or impulse noise may be present. Then, the MSE is no longer the ML function. In these situations, it is worth to apply a more robust cost function, as the Huber Robust cost function [40]. Using the SRM Induction principle, a form of the Huber cost function is applied in a natural way.

The regularization provided by the SRM Principle can be applied here [30, 78] in a similar way as in Support Vector Classifiers. The information taken from the data is, obviously, in the parameters $\boldsymbol{w}, b$. These data will be limited, usually only small data sets are available. If the training attempts to extract too much information from the data, the system will show overfitting, and attempting to extract low information will lead to an undertrained machine.

So, we can formulate an heuristic to traduce this trade-off into a functional:

> *Find the linear machine which approximates the correspondence function between $\boldsymbol{x}$ and y using the minimum possible energy from the data.*

As we have seen in Appendix B, the parameters $\boldsymbol{w}$ resulting of the application of the SRM is a linear combination of the input data, which has the minimum possible norm. Using the SVM will lead to a direct interpretation of the above heuristic. Following it results in the minimization of the primal functional

$$L_p = \frac{1}{2}||\boldsymbol{w}||^2 + \sum_{i=1}^{N} \ell(\xi_i, \xi_i') \tag{C.2}$$

subject to the constraints

$$\begin{aligned} y_i - \boldsymbol{w}^T \boldsymbol{x}_i - b &\leq \xi_i + \varepsilon \\ -y_i + \boldsymbol{w}^T \boldsymbol{x}_i + b &\leq \xi_i' + \varepsilon \\ \xi_i, \xi_i' &\geq 0 \end{aligned} \tag{C.3}$$

where the first expression is to be applied for positive errors and the second one for negative errors. Again, slack variables ξ_i, ξ_i' are forced to be nonnegative. Obviously, for each sample $\boldsymbol{x}_i$ only one of the slack variables will be different from zero. Parameter ε is used here to discard from the solution those samples whose output error are less that ε. In other words, we allow the positiveness of the conditions for estimation errors less than ε. That way, following a KKT condition similar to (B.10), all Lagrange multipliers associated to samples with $\xi_i, \xi_i' < \varepsilon$ will be zero. The solution for the Support Vector Regressor will be sparse, as it is for the case of classification.

If we simply choose $\ell(\xi_i, \xi_i') = \xi_i + \xi_i'$ into (C.2), this is equivalent to apply the so-called Vapnik or ε-insensitive loss function, which is [30, 78]

$$L_\varepsilon(\xi) = \begin{cases} 0 & |\xi| < \varepsilon \\ |\xi| - \varepsilon & \text{otherwise} \end{cases} \tag{C.4}$$

C.2   ROBUST COST FUNCTION

The optimization of primal (C.2) leads to a quadratic form, but the problem may be again ill-conditioned, so we need to add a small identity to the dot product matrix as we previously

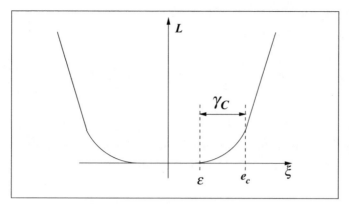

FIGURE C.1: Cost function with ε-insensitive, quadratic, and linear cost zones ($e_C = \varepsilon + \gamma C$)

did for the classification problem above. In order to find an interpretation for this small identity, we slightly modify the loss (C.4). The loss that we apply has the following expression [10]

$$L(\xi) = \begin{cases} 0 & |\xi| \leq \varepsilon \\ \frac{1}{2\gamma}(|\xi| - \varepsilon)^2 & \varepsilon \leq |\xi| \leq e_C \\ C(|\xi| - \varepsilon) - \frac{1}{2}\gamma C^2 & |\xi| \geq e_C \end{cases} \qquad (C.5)$$

where $e_C = \varepsilon + \gamma C$; ε is the insensitive parameter, and γ and C control the trade-off between the regularization and the losses. Three different regions allow to deal with different kinds of noise (see Fig. C.1): ε-insensitive zone ignores errors lower than ε; quadratic cost zone uses the L_2-norm of errors, which is appropriate for Gaussian noise; and linear cost zone limits the effect of outliers. Note that *(C.5)* represents Vapnik ε-insensitive cost function when γ is small enough, and Huber's [40] cost function when $\varepsilon = 0$. γ also plays the role of numerical regularization in the quadratic problem.

In most situations, the noise present at the signal output is the sum of thermal noise plus other contributions coming from external interference, which are usually sub-Gaussian. In these situations, the probability density function (PDF) is the convolution of both PDFs, but a good approximation near the origin is the Gaussian one. Then, a L_2 cost function will give the maximum-likelihood (ML) estimator. Far from the origin, the sub-Gaussian approach is the more convenient one. Then, a L_ρ, $\rho < 1$ (sub-linear) cost function will give the ML estimator. Nevertheless, to assure convergence, the cost function must be convex. The lowest degree cost function which is convex is $\rho = 1$, so the most convenient cost function far from the origin is L_1.

C.3   SOLVING THE FUNCTIONAL

The primal which fits the loss function *(C.5)* is

$$L_p = \frac{1}{2}\|\mathbf{w}\|^2 + \frac{1}{2\gamma}\sum_{i \in I_1}\left(\xi_i^2 + \xi_i'^2\right) + C\sum_{i \in I_2}\left(\xi_i + \xi_i' - \frac{\gamma C}{2}\right) \tag{C.6}$$

I_1, I_2 are the sets of samples for which losses are required to have a quadratic or a linear cost, respectively. These sets are not necessarily static during the optimization procedure. The constraints are the same as in *(C.3)*, which we rewrite as

$$\begin{aligned} -y_i + \mathbf{w}^T\mathbf{x}_i + b + \xi_i + \varepsilon &\geq 0 \\ y_i - \mathbf{w}^T\mathbf{x}_i - b + \xi_i' + \varepsilon &\geq 0 \\ \xi_i, \xi_i' &\geq 0 \end{aligned} \tag{C.7}$$

The corresponding primal-dual Lagrange formulation is

$$\begin{aligned} L_{pd} =& \frac{1}{2}\|\mathbf{w}\|^2 + \frac{1}{2\gamma}\sum_{i \in I_1}\left(\xi_i^2 + \xi_i'^2\right) + C\sum_{i \in I_2}\left(\xi_i + \xi_i' - \frac{\gamma C}{2}\right) \\ &+ \sum_{i=1}^{N}\alpha_i\left(y_i - \mathbf{w}^T\mathbf{x}_i - b - \xi_i - \varepsilon\right) \\ &+ \sum_{i=1}^{N}\alpha_i'\left(-y_i + \mathbf{w}^T\mathbf{x}_i + b - \xi_i - \varepsilon\right) \\ &- \sum_{i=1}^{N}\mu_i\xi_i - \sum_{i=1}^{N}\mu_i'\xi_i' \end{aligned} \tag{C.8}$$

The KKT conditions for the present problem are

$$\frac{\partial L_{pd}}{\partial \mathbf{w}} = 0$$

$$\frac{d L_{pd}}{d b} = 0 \tag{C.9}$$

$$\frac{\partial L_{pd}}{\partial \xi_i} = 0$$

$$\alpha_i, \alpha_i', \mu_i, \mu_i' \geq 0 \tag{C.10}$$

$$\alpha_i(-y_i + \mathbf{w}^T\mathbf{x}_i + b + \xi_i + \varepsilon) = 0$$

$$\alpha_i'(y_i - \mathbf{w}^T\mathbf{x}_i - b + \xi_i' + \varepsilon) = 0 \tag{C.11}$$

$$\mu_i\xi_i, \mu_i'\xi_i' = 0 \tag{C.12}$$

If condition (C.9) is applied, the following results hold

$$w = \sum_{i=1}^{N} (\alpha_i - \alpha_i') x_i \qquad (C.13)$$

$$\sum_{i=1}^{N} (\alpha_i - \alpha_i') = 0 \qquad (C.14)$$

$$\frac{1}{\gamma} \xi_i - \alpha_i - \mu_i = 0, \, i \in I_1$$

$$\frac{1}{\gamma} \xi_i' - \alpha_i' - \mu_i' = 0, \, i \in I_1 \qquad (C.15)$$

$$C - \alpha_i - \mu_i = 0, \, i \in I_2$$

$$C - \alpha_i' - \mu_i' = 0, \, i \in I_2 \qquad (C.16)$$

Result *(C.13)* is the solution for the parameters w. Also, we can see that, as $\xi_i > 0$ in intervals I_1 and I_2, then $\mu_i = 0$ in these intervals. From expressions *(C.15)* and *(C.16)*, the following equality holds for the Lagrange multipliers α_i (and, equivalently, for α_i')

$$\alpha_i = \begin{cases} C & \xi_i \geq \varepsilon + \gamma C \\ \frac{\xi_i}{\gamma} & \varepsilon \leq \xi_i \leq \varepsilon + \gamma C \\ 0 & 0 \leq \xi_i \leq \varepsilon \end{cases} \qquad (C.17)$$

which is just the derivative of the cost function *(C.5)*.

Now, we can apply expressions *(C.12)*, *(C.13)* and *(C.17)* to the primal-dual functional *(C.8)*, obtaining the dual

$$L_d = -\frac{1}{2} \sum_{i=1}^{N} \sum_{i=1}^{N} (\alpha_i - \alpha_i') x_i^T x_j (\alpha_i - \alpha_j')$$

$$+ \sum_{i=1}^{N} ((\alpha_i - \alpha_i') y_i - (\alpha_i + \alpha_i') \varepsilon) \qquad (C.18)$$

$$- \sum_{i \in I_1} \frac{\gamma}{2} (\alpha_i^2 + \alpha_i'^2) - \sum_{i \in I_2} \frac{\gamma C^2}{2}$$

Taking into account again that, in interval I_2, $\alpha_i, \alpha_i' = C$, we can change the last term of *(C.18)* by $\sum_{i \in I_2} \frac{\gamma}{2} (\alpha_i^2 + \alpha_i'^2)$. With that, the expression becomes

$$L_d = -\frac{1}{2} \sum_{i=1}^{N} \sum_{i=1}^{N} (\alpha_i - \alpha_i') x_i^T x_j (\alpha_j - \alpha_j')$$

$$+ \sum_{i=1}^{N} \left((\alpha_i - \alpha_i') y_i - (\alpha_i + \alpha_i') \varepsilon - \frac{\gamma}{2} (\alpha_i^2 + \alpha_i'^2) \right) \qquad (C.19)$$

A matrix expression of this equation can be easily found by rearranging terms and taking into account that $\alpha_i \alpha_i' = 0$

$$L_d = -\frac{1}{2}(\boldsymbol{\alpha} - \boldsymbol{\alpha}')^T [\boldsymbol{R} + \gamma \boldsymbol{I}](\boldsymbol{\alpha} - \boldsymbol{\alpha}') + (\boldsymbol{\alpha} - \boldsymbol{\alpha}')^T \boldsymbol{y} - (\boldsymbol{\alpha} + \boldsymbol{\alpha}')\mathbf{1}\varepsilon \qquad (C.20)$$

In this equation the numerical regularization term is γ, as we claimed at the beginning of the section. This parameter has an interpretation in terms of the applied cost function. Applying it is equivalent to apply a cost function which has quadratic interval between ε and $\varepsilon + \gamma C$. If the motivation for the use of γ is to raise the values of the eigenvectors of $\boldsymbol{R}$, then, its value is very low small, and the quadratic interval is negligible. Nevertheless, greater values of γ can also be used in order to make the interval wide enough to include all the data which is mainly corrupted by AWGN, letting the linear interval to the outliers, or samples corrupted by impulse or other sub-Gaussian noises.

Bibliography

[1] V. Vapnik, *The Nature of Statistical Learning Theory*, Springer–Verlag, NY, 1995.

[2] V. Vapnik, *Statistical Learning Theory, Adaptive and Learning Systems for Signal Processing, Communications, and Control*, John Wiley & Sons, 1998.

[3] N. Cristianini and J. Shawe-Taylor, *An Introduction To Support Vector Machines (and other kernel-based learning methods)*, Cambridge University Press, UK, 2000.

[4] K. R. Müller, S. Mika, G. Rätsch, K. Tsuda, and B. Schölkopf, "An introduction to kernel-based learning algorithms," *IEEE Transactions on Neural Networks*, vol. 12, no. 2, pp. 181–202, March 2001.doi:org/10.1109/72.914517

[5] J. Shawe-Taylor and N. Cristianini, *Kernel Methods for Pattern Analysis*, Cambridge University Press, 2004.

[6] T. Joachims, "Text categorization with support vector machines: learning with many relevant features," in *Proceedings of ECML-98, 10th European Conference on Machine Learning*, C. Nédellec and C. Rouveirol, Eds., Chemnitz, DE, 1998, no. 1398, pp. 137–142, Springer Verlag, Heidelberg, DE.

[7] S. Mukherjee, P. Tamayo, D. Slonim, A. Verri, T. Golub, J. Mesirov, and T. Poggio, "Support vector machine classification of microarray data," Tech. Report CBCL Paper 182/AI Memo 1676 MIT. 1999.

[8] S. LaConte, S. Strother, V. Cherkassky, J. Anderson, and X. Hu, "Support vector machines for temporal classification of block design fmri data," *Neuroimage*, vol. 26, pp. 317–329, March 2005.doi:org/10.1016/j.neuroimage.2005.01.048

[9] M. Martinez-Ramon and S. Posse, "Fast estimation of optimal resolution in classification spaces for fmri pattern recognition," *Neuroimage*, vol. 31, no. 3, pp. 1129–1141, July 2006.doi:org/10.1016/j.neuroimage.2006.01.022

[10] J. L. Rojo-Álvarez, G. Camps-Valls, M. Martínez-Ramón, A. Navia-Vázquez, and A. R. Figueiras-Vidal, "A Support Vector Framework for Linear Signal Processing," *Signal Processing*, vol. 85, no. 12, pp. 2316–2326, December 2005.doi:org/10.1016/j.sigpro.2004.12.015

[11] A. Ganapathiraju, J. Hamaker, and J. Picone, "Applications of support vector machines to speech recognition," *IEEE Transactions on Signal Processing*, vol. 52, no. 8, pp. 2348–2355, 2004.doi:org/10.1109/TSP.2004.831018

[12] J. Picone, "Signal modeling techniques in speech recognition," *IEEE Proceedings*, vol. 81, no. 9, pp. 1215–1247, 1993.doi:org/10.1109/5.237532

[13] M. Pontil and A. Verri, "Support vector machines for 3D object recognition," *IEEE Transactions on Pattern Analysis and Machine Intelligence*, vol. 6, pp. 637–646, 1998.

[14] K. I. Kim, M. O. Franz, and B. Schölkopf, "Iterative kernel principal component analysis for image modeling," *IEEE Transactions on Pattern Analysis and Machine Intelligence*, vol. 27, no. 9, pp. 1351–1366, 2005.doi:org/10.1109/TPAMI.2005.181

[15] D. Cremers, "Shape statistics in kernel space for variational image segmentation," *Pattern Recognition*, vol. 36, no. 9, pp. 1929–1943, 2003.doi:org/10.1016/S0031-3203(03)00056-6

[16] G. Camps-Valls and L. Bruzzone, "Kernel-based methods for hyperspectral image classification," *IEEE Transactions on Geoscience and Remote Sensing*, vol. 43, no. 6, pp. 1351–1362, June 2005.doi:org/10.1109/TGRS.2005.846154

[17] G. Gómez-Pérez, G. Camps-Valls, J. Gutiérrez, and J. Malo, "Perceptual adaptive insensitivity for support vector machine image coding," *IEEE Transactions on Neural Networks*, vol. 16, no. 6, pp. 1574–1581, November 2005.doi:org/10.1109/TNN.2005.857954

[18] G. Camps-Valls, L. Gomez-Chova, J. Muñoz-Marí, J. Vila-Francés, and J. Calpe-Maravilla, "Composite kernels for hyperspectral image classification," *IEEE Geoscience and Remote Sensing Letters,* vol. 3, no. 1, pp. 93–97, January 2006.doi:org/10.1109/LGRS.2005.857031

[19] G. Camps-Valls, J. L. Rojo-Álvarez, and M. Martínez-Ramón, Eds., *Kernel Methods in Bioengineering, Image and Signal Processing*, Idea Group Inc., 1996.

[20] M. J. Fernández-Getino García, J. L. Rojo-Álvarez, F. Alonso-Atienza, and M. Martínez-Ramón, "Support vector machines for robust channel estimation in OFDM," *IEEE Signal Processing Letters*, vol. 13, no. 7, pp. 397–400, July 2006.doi:org/10.1109/LSP.2006.871862

[21] S. Chen, A. K. Sanmigan, and L. Hanzo, "Support vector machine multiuser receiver for DS-CDMA signals in multipath channels," *Neural Networks*, vol. 12, no. 3, pp. 604–611, May 2001.

[22] S. Chen, A. K. Sanmigan, and L. Hanzo, "Adaptive multiuser receiver using a support vector machine technique," in *Proceedings of the IEEE Semiannual Vehicular Technology Conference, VTC 2001 Spring, CD-ROM*, Rhode, Greece, May 2001, pp. 604–608.

[23] S. Chen and L. Hanzo, "Block-adaptive kernel-based CDMA multiuser detector," in *Proceedings of the IEEE Conference on Communications*, NY, USA, 2002, pp. 682–686.

[24] S. Chen, S. Gunn, and C. J. Harris, "Decision feedback equalizer design using support vector machines," *IEE Proc. Vision, Image and Signal Processing*, vol. 147, no. 3, pp. 213–219, June 2000.

[25] G. N. Karystinos and D. A. Pados, "On overfitting, generalization, and randomly expanded training sets," *IEEE Transactions on Neural Networks*, vol. 11, no. 5, pp. 1050–1057, September 2000.doi:org/10.1109/72.870038

[26] G. S. Kimeldorf and G. Wahba, "Some results of Tchebycheffian Spline functions," *Journal of Mathematical Analysis and Applications*, vol. 33, 1971.

[27] A. N. Tikhonov and V. Y. Arsenen, *Solution to Ill-Posed Problems*, V. H. Winston & Sons, 1977.

[28] T. Poggio and F. Girosi, "Regularization algorithms for learning that are equivalent to multilayer networks," *Science*, vol. 247, 1990.

[29] C. J. C. Burges, "A tutorial on support vector machines for pattern recognition," *Data Mining and Knowledge Discovery*, vol. 2, no. 2, pp. 1–32, 1998.

[30] A. Smola, B. Schölkopf, and K. R. Müller, "General cost functions for support vector regression," in *Proceedings of the Ninth Australian Conference on Neural Networks*, Brisbane, Australia, 1998, pp. 79–83.

[31] D. P. Bersekas, *Constrained Optimization and Lagrange Multipler Methods*, Academic Press, 1982.

[32] J. Platt, "Fast training of support vector machines using sequential minimal optimization," in *Advances in Kernel Methods—Support Vector Learning*, B. Schölkopf, C. J. C. Burges, and A. J. Smola, Eds., MIT Press, Cambridge, MA, 1999, pp. 185–208.

[33] R. Courant and D. Hilbert, *Methods of Mathematical Phisics*, Interscience, 1953.

[34] M. A. Aizerman, É. M. Braverman, and L. I. Rozonoér, "Theoretical foundations of the potential function method in pattern recognition learning," *Automation and Remote Control*, vol. 25, pp. 821–837, 1964.

[35] J. L. Rojo-Álvarez, M. Martínez-Ramón, M. dePrado Cumplido, A. Artés-Rodríguez, and A. R. Figueiras-Vidal, "Support vector method for robust ARMA system identification," *IEEE Transactions on Signal Processing*, vol. 52, no. 1, pp. 155–164, January 2004.doi:org/10.1109/TSP.2003.820084

[36] J. L. Rojo-Álvarez, M. Martínez-Ramón, A. R. Figueiras-Vidal, A. García-Armada, and A. Artés-Rodríguez, "A robust support vector algorithm for nonparametric spectral analysis," *IEEE Signal Processing Letters*, vol. 10, no. 11, pp. 320–323, November 2003.doi:org/10.1109/LSP.2003.818866

[37] M. Martínez-Ramón, J. L. Rojo-Álvarez, G. Camps-Valls, A. Navia-Vázquez, E. Soria-Olivas, and A. R. Figueiras-Vidal, "Support vector machines for nonlinear kernel ARMA system identification," *IEEE Transactions on Neural Networks*, November, 2006 accepted for publication.

[38] S. L. Marple, *Digital Spectral Analysis with Applications*, Prentice-Hall, 1987.

[39] L. Ljung, *System Identification. Theory for the User*, 2nd edition, PTR Prentice Hall, Upper Saddle River, NJ, 1999.

[40] P. J. Huber, "Robust statistics: a review," Annals of Statistics, vol. 43, p. 1041, 1972.

[41] J. T. Kwok, "The evidence framework applied to support vector machines," *IEEE Transactions in Neural Networks*, vol. 11, no. 5, pp. 1162–1173, September 2000.

[42] J. T. Kwok and I. W. Tsang, "Linear dependency between ε and the input noise in ε-support vector regression," *IEEE Transactions in Neural Networks*, vol. 14, no. 3, pp. 544–553, May 2003.

[43] V. Cherkassky and Y. Ma, "Practical selection of SVM parameters and noise estimation for SVM regression," *Neural Networks*, vol. 17, no. 1, pp. 113–126, January 2004.doi:org/10.1016/S0893-6080(03)00169-2

[44] D. J. C. McKay, "Bayesian interpolation," *Neural Networks*, vol. 4, no. 3, pp. 415–447, May 1992.

[45] S. Haykin, *Adaptive Filter Theory*, 4th edition, Prentice-Hall, Englewood Cliffs, NJ, 2001.

[46] M. Martínez-Ramón, N. Xu, and C. Christodoulou, "Beamforming using support vector machines," *IEEE Antennas and Wireless Propagation Letters*, vol. 4, pp. 439–442, December 2005.

[47] C. C. Gaudes, J. Via, and I. Santamaría, "An IRWLS procedure for robust beamforming with sidelobe control," in *Sensor Array and Multichannel Signal Processing Workshop Proceedings*, July 2004, pp. 342–346.

[48] C. C. Gaudes, J. Via, and I. Santamaría, "Robust array beamforming with sidelobe control using support vector machines," in *IEEE 5th Workshop on Signal Processing Advances in Wireless Communications*, July 2004, pp. 258–262.

[49] M. Martínez-Ramón and C. Christodoulou, "Support vector array processing," in *Proceedings of the IEEE AP-S International Symposium*, 2006.

[50] M. Martínez-Ramón and C. Christodoulou, "Support vector array processing," *IEEE Transactions on Antennas and Propagation*, 2006, submitted for publication.

[51] A. El Zooghby, C. G. Christodoulou, and M. Georgiopoulos, "Neural network-based adaptive beamforming for one and two dimensional antenna arrays," *IEEE Transactions on Antennas and Propagation*, vol. 46, no. 12, pp. 1891–1893, 1988.doi:org/10.1109/8.743843

[52] M. Sánchez-Fernández, M. de Prado-Cumplido, J. Arenas-García, and F. Pérez-Cruz, "SVM multiregression for nonlinear channel estimation in multiple-input multiple-output systems," *IEEE Transactions on Signal Processing*, vol. 52, no. 8, pp. 2298–2307, August 2004.doi:org/10.1109/TSP.2004.831028

[53] B. Schölkopf, A. Smola, and K.-R. Müller, "Kernel principal component analysis," in Advances in Kernel Methods—SV Learning, B. Schölkopf, C. J. C. Burges, and A. J. Smola, Eds., MIT Press, Cambridge, MA, 1999, pp. 327–352.

[54] S. Verdú, "Multiuser detection," in *Advances in Statistical Signal Processing*, vol. 2, JAI Press, Greenwich, CT, 1993.

[55] L. Ljung, *System Identification. Theory for the User*, Prentice Hall, NJ, USA, 1987.

[56] R. Roy and T. Kailath, "ESPRIT-estimation of signal parameters via rotational invariance techniques," *IEEE Transactions on Acoustics, Speech and Signal Processing*, vol. 37, no. 7, July 1989.

[57] H. L. Van Trees, *Optimum Array Processing (Detection, Estimation, and Modulation Theory, Vol IV)*, Wiley Interscience, 2002.

[58] C. Christodoulou and M. Georgiopoulos, *Applications of Neural Networks in Electromagnetics*, Artech House Inc., Boston, MA, USA, 2001.

[59] M. Pastorino and A. Randazzo, "A smart antenna system for direction of arrival estimation based on a support vector regression," *IEEE Transactions on Antennas and Propagation*, vol. 53, no. 7, pp. 2161–2168, July 2005.doi:org/10.1109/TAP.2005.850735

[60] J. A. Rohwer and C. T. Abdallah, "One-vs-one multiclass least squares support vector machines for direction of arrival estimation," *Applied Computational Electromagnetics Society Journal*, vol. 18, no. 2, pp. 34–45, July 2003.

[61] J. A. Rohwer, C. T. Abdallah, and C.G. Christodoulou, "Least squares support vector machines for direction of arrival estimation," *Antennas and Propagation Society International Symposium*, vol. 1, pp. 57–60, June 2003.

[62] J. C. Platt, N. Christianini, and J. Shawe-Taylor, "Large margin DAGs for multiclass classification," in *Advances in Neural Information Processing Systems*, MIT Press, Cambridge, MA, 2000, pp. 547–553.

[63] Q. Zhao and J. C. Principe, "Support vector machines for SAR automatic target recognition," *IEEE Transactions on Aerospace and Electronics Systems*, vol. 37, no. 2, pp. 643–654, April 2001.

[64] B. Krishnapuram, J. Sichina, and L. Carin, "Physics-based detection of targets in SAR imagery using support vector machines," *IEEE Sensors Journal*, vol. 3, no. 2, pp. 147–157, April 2003.doi:org/10.1109/JSEN.2002.805552

[65] A. Massa, A. Boni, and M. Donelli, "A classification approach based on SVM for electromagnetic subsurface sensing," *IEEE Transactions on Geoscience and Remote Sensing*, vol. 43, pp. 1–10, May 2005.

[66] A. N. Srivastava, N. C. Oza, and J. Stroeve, "Virtual sensors: using data mining techniques to effiecintly estimate remote sensing spectra," *IEEE Transactions on Geoscience and Remote Sensing*, vol. 43, no. 3, pp. 1–10, March 2005.

[67] S. Poyhonen, M. Negrea, A. Arkkio, H. Hyotyniemi, and H. Koivo, "Fault diagnosis of an electrical machine with multiple support vector classifiers," in *Proceedings of the 2002 IEEE International Symposium on Intelligent Control*, pp. 373–378, October 2002.

[68] E. Bermani, A. Boni, S. Caorsi, and A. Massa, "An innovative real-time technique for buried object detection," *IEEE Transactions on Geoscience and Remote Sensing*, vol. 41, no. 4, pp. 927–931, April 2003.doi:org/10.1109/TGRS.2003.810928

[69] S. Caorsi, D. Anguita, E. Bermani, A Boni, M. Donelli, and A Massa, "A comparative study of NN and SVM-based electromagnetic inverse scattering approaches to online detection of buried objects," *Applied Computational Electromagnetics Society Journal*, vol. 18, no. 2, pp. 1–11, July 2003.

[70] C. Gaudes, J. Via, and I. Santamaria, "Robust array beamforming with sidelobe control using support vector machines," in *Fifth IEEE Workshop on Signal Processing Advances in Wireless Communications*, Lisboa, Portugal, July 2004.

[71] J. Li, P. Stoica, and Z. Wang, "On the robust Capon beamforming and diagonal loading," *IEEE Transactions on Signal Processing*, vol. 51, no. 7, pp. 1702–1715, July 2003.doi:org/10.1109/TSP.2003.812831

[72] A. Mirzai, C. F. Cowan, and T. M. Crawford, "Intelligent alignment of waveguide filters using a machine learning approach," *IEEE Transactions on Microwave Theory and Techniques*, vol. 37, no. 1, pp. 166–173, January 1989.doi:org/10.1109/22.20035

[73] V. Vapnik, *Estimation of Dependences Based on Empirical Data*, Nauka, Moscow, 1979.

[74] C. Cortes and V. Vapnik, "Support vector networks," *Machine Learning*, vol. 20, pp. 273–97, 1995.

[75] D. Luenberger, *Linear and Nonlinear Programming*, 2nd edition, Springer, 2003.

[76] F. Pérez-Cruz, P. L. Alarcón-Diana, A. Navia-Vázquez, and A. Artés-Rodríguez, "Fast training of support vector classifiers," in *Advances in Neural Information Processing Systems 13*, T. K. Leen, T. G. Dietterich, and V. Tresp, Eds., MIT Press, 2001.

[77] S. Verdú, *Multiuser Detection*, Cambridge University Press, 1998.

[78] A. Smola and B. Schölkopf, "A tutorial on support vector regression," NeuroCOLT Technical Report NC-TR-98-030, Royal Holloway College, University of London, UK, 1988.

Index